Elias Jackson "Lucky" Baldwin

California Visionary

by

Sandra Lee Snider

The Stairwell Group
Los Angeles

The Stairwell Group, Publishers
14400 Mulholland Drive
Los Angeles, California 90077

Printed in the United States of America
Library of Congress Catalog Card Number: 87-62917
ISBN 0-9619623-0-5
First Printing

Contents

Acknowledgments

My thanks to Carla Hind, whose initial enthusiasm inspired both this author and the publishers to set forth on the endeavor to bring the story of Lucky Baldwin to a broader audience.

The vision that was Baldwin's a hundred years ago has become today's reality in the city of Arcadia and its racetrack, and in the old Baldwin homesite itself which has been preserved on the grounds of today's Los Angeles State and County Arboretum.

The author, Historical Curator at the Arboretum, acknowledges that institution's support. The photographs in the book are from the Arboretum archives, supplemented with photos borrowed from the Huntington Library, Big Bear Historical Society, Arcadia Public Library, and private donors such as Jack McCaskill, Julian Fisher, Elizabeth Cleminson, Vesta Tucker Reevers, and members of the Baldwin family.

My personal appreciation is extended to friends who donated their time and thoughtful consideration in preliminary readings and decision making.

Rancho Santa Anita
Chain of Title
King of Spain-1771
Mexican Nation-1822
Hugo Reid-1839
Henry Dalton-1847
Joseph A. Rowe-1854
Wm. Corbitt and A. Dibblee-1858
Wm. Wolfskill-1865
Harris Newmark-1872
E. J. "Lucky" Baldwin-1875
Harry Chandler-1936
State of California-1947
CALIFORNIA REPUBLIC
CALIFORNIA REPUBLIC

Introduction

Drastic and accelerating change was on the Southern California horizon in 1875, and Elias Jackson Baldwin, a new owner of extensive Southland acreage, both embraced and reflected such change. The days of baronial cattle ranching were over by the late 1860s, replaced by a more intensified agricultural development of the land and an increasing urbanization. Small farms and fledgling cities were soon to flourish in the bloom of dynamic economic growth, the atmosphere heady with the largess of transcontinental railroad connections and escalating property values. Subdivision nibbled at the edges of the large ranches during the boom years of the eighties, until an inevitable "bust" slowed the process.

Land development correspondingly languished in the following decade. Paper cities crumbled, small property owners lost investments, and large landowners such as Baldwin were forced to acknowledge the irony of the term "land poor." Rich in property, but often lacking the cash to meet fixed charges, the land poor mortgaged their holdings to pay their bills and in turn accepted a lifetime of interest payments on their loans.

The turn of the century saw an improving national economy and new hope for those owners who had survived the rigors of a decade of depression. In Southern California, suburbanization rode the rails of the booming Pacific Electric, sparking new life and glimmers of profit for mortgage-encrusted properties. A renewed nibbling of landed estates ensued, this time the orange groves and vineyards joining the lesser subdivisions on the auction block. Ranches became cities with increasing regularity, and by the second decade of the century the large landowner was either an anachronism or a burial statistic. The stories of the great ranches were already becoming history.

Elias Jackson "Lucky" Baldwin was a typical practitioner of the art of aggrandizement and survival through boom and bust, through mortgage and mortgage foreclosure. With luck, with skill, enthusiasm and perseverance, Lucky Baldwin created and maintained a showcase ranch that for thirty-five years was the envy of many. With equal and timely enthusiasm, he subdivided ranch lands, launched townsites, even fathered a city. And he did so while teetering on the brink of insolvency. A detailed examination of the management, methods, and manipulations of Elias Baldwin affords both an adventurous tale and an illustrative analysis of Southern California enterprise during a significant period of the area's growth.

An investigation of available references reveals a cache of twentieth century gossip, reminiscenses, and anecdotes about Lucky Baldwin and his ranch, many of them injudiciously lent the credibility of the printed word by writers eager to romanticize a colorful personality. In 1945, MGM Studios prepared a "Lucky Baldwin" movie script that, despite the prospect of Clark Gable in the lead, was never made. Even *True* ("the Man's magazine") found Lucky Baldwin pulp for their mill under the billing "Millionaire, King of a Thousand Queens" (February, 1957).

Anecdote and reminiscence were put to more solid use in C.B. Glasscock's biography (1933) of Elias Baldwin. The only such undertaking to date, *Lucky Baldwin: The Story of an Unconventional Success*, has long served, for lack of a better, as the most comprehensive account of the Baldwin career. The annoying lack of documentation from a biographer who claimed to have consulted numerous newspapers, plus other references, is at least ameliorated by clear source citations of oral histories. Mr. Glasscock began his research while a number of Baldwin contemporaries were still alive, and the preservation of their stories is a credit to the author.

Anecdotal primary sources are equally plentiful, and their quality equally varied. "If you will look in Bancroft's 'Chronicles of the Builders' you will find all you want to know about me," Lucky Baldwin told an interviewer in 1907.[1] The personal anecdotes and sequences of events that E.J. related to Bancroft in 1891-92 were repeated with sometimes significant variations to other interviewers over the years, but if one accepts Lucky Baldwin's latter-day judgment on the merits of *Chronicles of the Builders* and can filter the obvious discrepancies, the recorded Baldwin reminiscences provide an interesting research tool. Of greater value are the ranch statistics and descriptions gathered by Bancroft's researchers, at least one of whom paid a personal visit to the ranch to record his observations. In the absence of actual ranch records, such first-hand data-gathering is an important link in re-creating not only the mood, but the mechanics of the past.

"The Baldwin family records would have been excellent historical records as they were complete—payrolls, checks, invoices, etc.—for two generations," wrote Ray Knisley, former manager of the Anita Baldwin ranch, to an earlier researcher.[2] According to Lowry McCaslin (Mr. Knisley's predecessor), Lucky Baldwin's daughter, Anita, stored the documents in a vault under the gymnasium at her Anoakia homesite. After the sale of the last of the Santa Anita ranch lands in 1936 and Anita's subsequent death three years later, her heirs requested that the contents of the vault be destroyed.[3] "I personally supervised their destruction," wrote Ray Knisley. "They were burned."[4]

The irretrievable loss of such documentation provides a challenge to the researcher seeking to examine the workings of the Baldwin Ranch. Happily, though inexplicably, a few random ranch and estate records were available at the Huntington Library, donated by a granddaughter of Anita Baldwin. Beyond their intrinsic value, clues in these documents led to searches at the County Hall of Records, while other avenues of approach afforded significant dates around which to examine period news reports. The methodology of town and city-making in particular unfolds in appreciable detail both in the advertising and in the local news sections of the area's newspapers.

A patchwork of sources has been assembled, with scraps from Bancroft, Glasscock, and others, from preserved pamphlets and brochures, from maps, deeds, court cases, and other surviving documents, from a large number of newspaper articles, from taped reminiscences and maps drawn from memory by oldsters who were youngsters during the prime of the Baldwin Ranch. From the scraps of information, a pattern emerges of an extraordinary individual both molding and being molded by dynamic forces of change in a Southland that was his adopted, but clearly preferred home.

Elias Jackson Baldwin was born April 3, 1828, in Hamilton County, Ohio, the fourth of five children. Removing to northwestern Indiana (near South Bend) in 1834, young Elias grew up in a farming family, working the soil and acquiring at an early age the love and knowledge of land that would work to his advantage in later years.[5] A year at Wabash College in Crawfordsville completed his formal education,[6] and shortly thereafter the headstrong Elias and neighbor Sarah Ann Unruh, both teenagers, eloped and were married in Berrien County, Michigan. Back home, the newlyweds farmed and traded in horses, accumulating $2000 in savings that enabled them to establish a grocery store in Valparaiso, Indiana. It was there that daughter Clara was born in 1847.[7]

Greener pastures were found in New Buffalo, Indiana, where the Baldwins opened a general store and hotel, later enlarging their operations to include grain trading on the new Illinois Canal.[8] Racine, Wisconsin, was the last of the couple's eastern moves, and it was from there that Elias, Sarah Ann, and Clara, in 1853, began their wagon train journey to the West. With six other Racine residents, Baldwin fitted out four wagons for the trek, two of them loaded with brandy, tobacco, and tea for trading. At Council Bluffs, they joined a larger train and proceeded without undue incident to Salt Lake, where a mercantile firm purchased Baldwin's tobacco and tea, and the brother of Brigham Young selected the brandy at $16 a gallon.[9] Other men jettisoned family heirlooms while crossing the plains; Elias Baldwin claimed a $3000 to $4000 profit on his Salt Lake sales, then reinvested part of his earnings in horses that he later sold to Sacramento buyers at a 400 percent markup.[10]

The Baldwin family more than doubled its Wisconsin capital while crossing the plains, and in San Francisco, E.J. used $5000 of his bankroll to buy a hotel.[11] A series of business and real estate investments ensued, each a modest, but profitable undertaking. Baldwin even enjoyed success in the brick business, though he claimed that when he first entered a partnership with an established owner, he "hardly knew a brick from a stone." He learned, and soon began his own business with a lucrative government contract to supply bricks for Fort Point and Alcatraz.[12] "If a man is determined, he can do anything in the world," claimed entrepreneur Baldwin.[13]

Determination could not, however, salvage his personal life. E.J. and Sarah Ann lost two infants, a son and a daughter, born in San Francisco, and in 1862 Mr. and Mrs. Baldwin were divorced.[14] Elias left San Francisco for Virginia City and the excitement of the Comstock. While running a lumber yard there, E.J. studied the mines and invested cautiously, his transactions "rather in the nature of investments than of speculation."[15] Constantly supplementing his own knowledge with reports from a corps of mining engineers kept on his payroll, Baldwin came to be known as a shrewd and well-informed operator on the San Francisco Stock Ex-

Elias Jackson Baldwin, Age 47, 1875

change.[16] "Lucky" was soon to be added to the description.

In 1867, with profits from the sale of several stocks, E.J. treated himself to a world tour. He locked his depressed Hale and Norcross shares in a safe, with orders to his broker to sell if they should reach their $800 a foot purchase price.[17] The price was reached and rose precipitously. When Baldwin returned to San Francisco in 1869, he brought along a new wife, Mary Cochrane, plus what became a good luck charm—the key to his safe.[18] Unable to retrieve his client's stocks without a key, E.J.'s broker had watched Hale and Norcross soar to $12,000 a foot by 1869, and that was the price at which the returning Baldwin was able to sell.[19] Though E.J. held only a small interest in Hale and Norcross, the story of his "luck" spread to the press, where the sobriquet "Lucky" Baldwin was popularized. Elias Jackson Baldwin resented the implications. When asked years later how he had gotten the nickname, Lucky replied, "That's a hard one. I've worked hard for everything I've gotten in life."[20]

E.J. continued to work hard in Comstock investments. In 1870, he sold his Crown Point shares for $2.5 million, later his Best and Belcher, purchased at $2 a share, for 750 times the purchase price.[21] Baldwin was less lucky in his next investment, however. Hesitating to compete with Mackay, Fair, Flood, and O'Brien, who were buying undeveloped land adjacent to Best and Belcher, E.J. instead bought substantially into the Ophir mine. In 1873, the Silver Kings brought in the "Big Bonanza" on their newly developed Consolidated Virginia land, and Baldwin must have cringed at the opportunity he had missed.[22] Both ego and pocketbook were assuaged, though, as, with Ophir stock moving up in price in the wake of the Bonanza, E.J. was able to outmaneuver Ralston and the "bank ring" to gain control of over half of the 108,000 mine shares.[23] In December, 1874, Ophir reached its highest price at $350 a share, and Lucky Baldwin sold to Ralston for cash plus a note for a $3.6 million balance.[24] "He retired from the fray with $5 million," according to Bancroft.[25]

Chapter One

Land Acquisition and Initial Development

The year 1875 was a pivotal one for Lucky Baldwin—and for Los Angeles County. E.J. and his $5 million did not slip quietly into the Southland. Readers of the *Los Angeles Express* must have chuckled upon reading that the San Francisco millionaire had dropped his fortune on a local city street, where it was retrieved but not immediately returned by E.J.'s prankster brother-in-law, Tom Fawcett. Baldwin's errant pocketbook, explained the *Express*, "contained two certificates of deposit, unendorsed, for $800,000 each, and a note signed by Ralston, Sharon, and others of that ilk for $3.6 million."[1] A separate "Local Item" dropped the other shoe. "Lucky Baldwin has made up his mind to come to Los Angeles and live."[2]

Elias Baldwin was not a stranger to the Southland in 1875. A year and a half earlier, rumors of gold in the San Bernardino Mountains near Bear Valley had prompted mining purchases by two Baldwin associates, Samuel Baird and Samuel Curtis. Curtis, Baldwin's manager at the Ophir Mine, was named principal owner in January, 1874, when Gold Mountain Mining Company was incorporated, though indications were that E.J. owned substantial interest in the operation.[3] In February, 1875, the *Express*, in fact, identified Baldwin as "one of the chief owners of the Gold Mountain Mine,"[4] and in retrospect Bancroft explained that Baldwin's mining interests were "at first held by a joint-stock company, with himself as principal stockholder, but a few years afterward he became sole proprietor."[5]

E.J. embarked on his Gold Mountain venture with customary vigor. Some $250,000 was expended on facilities alone, and 180 miners were put on the payroll.[6] A forty-stamp mill was erected, the largest in Southern California, and on March 8, 1875, Baldwin and a party of friends attended opening day ceremonies for the engineering marvel.[7] Lucky spent two weeks at the mine, then departed triumphant carrying a large ore sample he had personally chipped from the ledges. Triumph, however, was short-lived. In May, the Gold Mountain mill shut down twenty of its forty stamps, and on October 11, 1875, it closed completely.[8] Baldwin doggedly held the property, plus some six thousand acres surrounding it, but hopes for a profitable investment were turned from Bear Valley to Los Angeles County, where the beauty and fertile soil of Rancho Santa Anita had already caught farmer Baldwin's eye.

Gold Mountain Mine, Big Bear

"I came down to look at a mine, but when I

saw this ranch there was nothing that would make me happy but to own it," reminisced Baldwin.[9] Santa Anita was owned by Harris Newmark's company in 1875 and, as E.J. remarked, the two men "had a great deal of dickering" over it.[10] Baldwin offered Newmark $150,000 for acreage purchased by the company three years earlier for $85,000. "Learning that we wanted two hundred thousand dollars, he started off in a huff," remarked Newmark. "Then Reuben Lloyd, the famous San Francisco attorney who accompanied him, said on reaching the sidewalk, 'Lucky, go back and buy that ranch or they'll raise the price on you!' and Baldwin returned."[11] The purchase, Baldwin's first (excluding Bear Valley) in the Southland, was completed March 19, 1875, "the sum paid," reported the newspapers, "being $200,000, cash in hand."[12]

The purchase price was verified in a deed filed the following day, as E.J., through brother-in-law Tom Fawcett,[13] acquired what was described as eight thousand acres of Rancho Santa Anita—bounded on the west by the property of A.B. Chapman (roughly today's Michillinda Avenue) and on the east by the Azusa de Duart Rancho—plus 432 acres made up of fractions of sections of land outside the original Santa Anita Rancho, near the mouth of Santa Anita Canyon.[14] All water rights in Santa Anita Canyon were also included in the sale.[15] "This is the largest real estate transaction ever recorded in Los Angeles County," enthused the *Los Angeles Express.*

> We have no hesitation in saying that the sale of the Santa Anita has added of itself 10 per cent to the property valuation of Los Angeles County, and it is but an earnest of sales that will come fast upon its heels.[16]

The *Herald* was equally pleased.

> This is one of the most attractive pieces of ranch property in the State, and for few has nature done so much. The tract embraces every variety of soil and scenery . . . and is watered with natural streams and dotted here and there with pretty natural lakes. The place is capable of being converted into a perfect paradise, and Mr. Baldwin's taste, liberality, and abundant means will be immediately applied to this object. It is his intention to make this place the most elegant on the coast, with deer parks, trout ponds, a drive, a mile in length, through an avenue of orange trees, and the hill that commands a complete view of the domain and the surrounding country, has been fixed upon as the site for an elegant

Baldwin Ranch—Indian Hut, 1878

> dwelling. It is the intention of the purchaser to retain as far as possible the natural beauties and curiosities of the place, even to the old adobe house and Indian huts, and by his new improvements make a happy blending of the works of nature and art.[17]

Neither the *Herald* nor the *Express* had any doubt that Baldwin and his millions would seek out further Southland property. Mr. Baldwin, reported the *Express*, "has no hesitation in telling his friends that if the figures are not made too steep he will put nearly all of that $3.6 million loose cash in real estate here."[18] Adjoining Santa Anita on the south was Rancho San Francisquito, and within days of Baldwin's original purchase, rumors were circulating of negotiations between E.J. and San Francisquito owner, Louis Wolfskill. Wolfskill had acquired what was referred to as the Westerly Two-Thirds of San Francisquito as security on a loan to his father-in-law, Henry Dalton,[19] and on October 13, 1875, a total of 3,695 acres of the property was deeded to E.J. Baldwin by Wolfskill, F.P.F. Temple, and William Workman for $50,000.[20]

How Temple and Workman acquired an interest in the property is not known, and also mysterious is the fact that a map of the "Western Two-Thirds of Rancho San Francisquito," recorded in 1891 "at the request of E.J. Baldwin," described the property "surveyed by John Goldsworthy, 1869, for sale by Lewis Wolfskill," as containing an inflated 5,929 acres.[21] Glasscock, without documentation, states that Baldwin originally purchased "several hundred" acres from Louis Wolfskill for $10,000 and only later added "the adjoining acreage of San Francisquito."[22] If Glasscock is correct, E.J., then, acquired the total acreage in two separate purchases, paying a total of $60,000. The surmise, in fact, is supported by court testimony (in a 1912 estate suit by a daughter of Henry Dalton) that Dalton gave Louis Wolfskill permission to sell the San Francisquito lands "with the understanding that out of the proceeds each would receive $30,000 for his equity."[23]

Whatever the means of purchase, by early 1876 a visiting *Herald* reporter noted that the Baldwin Ranch now contained fifteen thousand acres. "Mr. Baldwin has owned this ranch less than a year," continued the newsman, "and in this short time has made many improvements."[24] Baldwin proposed spending $100,000 to distribute cienega (spring) and stream waters over his ranch, and orders for the engineering of wells, reservoirs, and irrigation systems were issued to superintendents who oversaw operations during E.J.'s frequent business absences.[25] Land was cleared and prepared for tillage, protected by miles of fencing. Within three years, twelve hundred acres had been set out in fruit and nut trees, three hundred acres in vineyard, and there were extensive nurseries to hold future stock.[26] "Mr. Baldwin has spent $30,000 perfecting his facilities for the manufacture of wines," reported the *Express*, "and has one of the most complete establishments in the state."[27]

An estimated two hundred men were continuously employed in the first years of development of the Baldwin Ranch. Added to the planting, irrigating, and fencing projects was an extensive program of construction and beautification. Some thirty buildings were erected, including stables, barns, storehouses, several "Swiss cottages" (presumably for employee families), a boarding house, general store, a school for ranch children, and, for Baldwin's newly purchased thoroughbred racing stock, separate stables and a private training track.[28] Specimen trees "from Europe, Japan, China, South America, and the West Indies" were imported and set out around the homesite. "The excellent taste of the proprietor, with an expenditure of at least four hundred thousand dollars," wrote a contemporary, "is fast transforming the primitive elements of beauty into an earthly paradise."[29]

Baldwin Adobe and Lake, 1878

"I believe that Baldwin's ranch will become the best improved, most valuable and noted in this whole valley," said a *Los Angeles Herald* reporter.[30] The words were not only prophetic, but the ranch was to become the largest in the valley as well when the suspension of the Temple and Workman Bank on September 1, 1875, offered Lucky Baldwin an unprecedented opportunity for aggrandizement. The Temple and Workman Bank in Los Angeles, jointly owned by F.P.F. Temple and his father-in-law, William Workman, had never had a strong financial reputation. "Mr. Temple's only qualification in a borrower was that he must be poor," Isaias Hellman had declared upon dissolving an earlier banking partnership with Temple.[31] In the wake of the 1875 Bank of California panic, both Hellman's (and John Downey's) Farmers and Merchants Bank and the Temple and Workman Bank closed their doors, but Hellman confidently reopened on October 1, while Temple continued a frantic search for funds.[32] Elias Baldwin, reputedly "third in the list of wealthy men on the Pacific coast,"[33] and certainly concerned about the financial situation in Southern California, agreed to tender a loan.

On December 2, 1875, Temple, Workman, and their close friend, Juan Sanchez, executed a promissory note in the sum of $310,225 to Elias Jackson Baldwin, payable two years after the date with interest at the rate of one percent per month, payable monthly, and if delinquent, two percent, compounded, for each unpaid month.[34] Security for the loan was a blanket mortgage conveyed to Baldwin on the following lands:[35]

1) eighty acres located southeast of the city of Los Angeles (between Fifty-fifth and Fifty-eighth Streets, Main Street, and Park Avenue, later known as the "Lucky Baldwin Tract");

2) Rancho Potrero Grande (4,431 acres), as described in the patent issued to Juan Sanchez

Map of Old Spanish and Mexican Ranchos

in 1859, excepting land previously deeded to Margarita Workman de Temple by her father;[36]

3) that portion of Rancho La Puente conveyed to William Workman by John Rowland in 1868, excepting lands previously deeded to Margarita Workman de Temple, E.L. and Robert Beale, Frederick Lambourne, Joseph Workman, and the Southern Pacific Railroad (after exceptions, 17,964 acres);

4) Rancho Potrero de Felipe Lugo (2,042 acres);[37]

5) Rancho La Merced (2,363 acres), as described in the patent issued to Juan Sanchez and F.P.F. Temple in 1872;

6) the unsold portion of Temple and Workman's land parcel known as Block One, in Los Angeles (bounded by First, Second, Spring, and Main Streets);

7) 240 feet on Spring Street, Los Angeles; and

8) the Temple Block in Los Angeles.

Temple's interest in the Rancho La Cienega o Paso de la Tijera was not part of the mortgage conveyed to Lucky Baldwin, but evidence presented by A.J. Hutchinson indicates that E.J.'s acquisition of the property was an unofficial condition for the loan. Temple, Hutchinson, Daniel Freeman, and Henry Ledyard (cashier at the Temple and Workman Bank) each owned a quarter interest in La Cienega, and each had signed an agreement not to sell his interest without first offering it to one of the other three owners at the same price.[38] In negotiating the bank loan, according to Hutchinson, Baldwin insisted that he be sold Temple and Ledyard's half of Rancho La Cienega, and at the price originally paid for it. Hutchinson objected and offered the same price, but was refused "as Baldwin had said that unless he got this Cienega interest he would not lend them one cent."[39] On December 2, 1875, the same day the mortgage was signed, Temple and Ledyard deeded an undivided one half of the 3,317 acre Rancho La Cienega to E.J. Baldwin for $35,000.[40] In 1886, Arthur Hutchinson deeded the remaining half interest to Baldwin (Hutchinson had apparently acquired Freeman's quarter), plus an adjoining 360 acres, for $60,000.[41]

The Temple and Workman Bank reopened December 6, 1875, but the long suspension, "the well-known terms of the Baldwin loan, and the remembered looseness of Temple's banking policies fostered a lingering distrust among many citizens."[42] The borrowed money dwindled, and on January 7, 1876, Baldwin loaned the bank an additional $30,000, the second note secured by Temple and Workman's 4,246 sheep (including all wool and progeny), plus a $12,000 claim held by the bankers against the federal government.[43] It was to no avail. The Temple and Workman Bank crashed, and on January 13, 1876, its business and the assets of F.P.F. Temple and William Workman were assigned for settlement to E.F. Spence and Daniel Freeman. With creditors knocking on the doors, an already ailing Workman committed suicide May 17, 1876; Temple shortly thereafter suffered a paralytic stroke.[44]

On March 5, 1877, Baldwin presented a claim for the promissory notes plus interest, $449,376, which was then past due and upon which no payment had ever been made.[45] Foreclosure was allowed, though another two years were spent in court as several defendants brought suit to contest E.J.'s mortgages. On February 13, 1879, when Baldwin filed a second decree stating that the amount now due on the notes was $576,574, it was ordered by the court that the mortgaged land be sold at public auction "to pay Plaintiff the amount due, with costs and interest."[46]

As the holder of the mortgage, Lucky Baldwin was "entitled to first payment of the proceeds of the sale of land," and at the auction of September 20, 1879, Camilo Martin (acting for Lucky Baldwin)[47] entered bids sufficient to pay the Temple-Workman indebtedness. The only exceptions to Baldwin's acquisition of the mortgaged lands were John Jones' lien on Block One, which was found superior to Baldwin's, and Alexander Weill's lien (later superseded by Newmark and Company) on the Temple Block, also found superior to Baldwin's.[48] Completing the demolition of assets, further decree awarded E.J. the multiplying Temple/Workman sheep, their wool, and the $12,000 government claim which had been held as collateral on the second ($30,000) loan. [49] Finally, for the impoverished Temple, Workman, and Sanchez families, Baldwin circumspectly provided for retention of their old family homesteads on the mortgaged lands. To the widow Temple, E.J. deeded fifty acres and the "Temple Homestead" at La Merced;[50] to William Workman's grandson, he arranged title to the "La Puente Homestead" and seventy-five acres;[51] and to Sanchez, he awarded an apparent lifetime (only) use of the family adobe at La Merced plus two hundred acres (when Sanchez's widow died in 1892, Baldwin "filed an action and received the adobe and the land from the Sanchez children").[52]

By 1880, five years after his Southland debut, Lucky Baldwin was the owner of a substantial amount of real estate. The Ranchos Santa Anita, San Francisquito, La Cienega, half of La Puente, Potrero Grande, Potrero de Felipe Lugo, and La Merced were his, as was Rancho Potrero Chico (Potrero de la Mission Vieja de San Gabriel), 19.33 acres deeded July 28, 1875, from John Downey (as guardian of W. and L. Cardwell) to Tom Fawcett, for $1,100.[53] In Los Angeles, too, E.J. owned the 240 feet on Spring Street, from which he received a steady rental income,[54] and the eighty-acre Lucky Baldwin Tract. Total Baldwin holdings, at least those verified, were just short of 45,000 acres.

That Baldwin owned other Los Angeles County lands is a not unbelievable supposition, one propounded, in fact, by numerous writers. Baldwin himself told an interviewer, "I have seven ranches together, in all 56,000 acres."[55] Bancroft reported 54,000 acres in the Baldwin Ranch;[56] and a Baldwin land sales pamphlet mentioned "50,000 acres in the heart of the San Gabriel Valley."[57] Walter Lindley, in *California of the South* (1888), the only (discovered) source to enumerate the acreage, listed the Baldwin holdings as: Santa Anita, 10,000 acres; La Puente, 19,000 acres; San Francisquito, 6,000 acres; Felipe Lugo, 3,000 acres; Potrero Grande, 5,000 acres; La Merced, 3,000 acres; Potrero Chico, 100 acres; La Cienega, 4,000 acres; and "4,000 acres of land in other parts of Los Angeles."[58] Lindley's figures, it should be noted, tend to be rounded upward to the nearest thousand (or more), and it is only by adding the mysterious 4,000 acres "in other parts of Los Angeles" that he totalled 54,000 acres.

The evidence of Baldwin holdings in the 50,000 plus range is plausible, however, though only circumstantial. Although no ranches beyond those mentioned were among the inventoried estate of Lucky Baldwin at the time of his death,[59] the possibility exists that over the course of Baldwin ownership, acreage may have been quietly added to any or all of the extant ranches. In a taped reminiscence, a former employee, ninety-eight year old Samuel Butler, indicated that Baldwin indeed added land to his ranches, supplementing his program of acquisition with a liberal interpretation of public land laws. Because one man could file on only limited amounts of public land, Baldwin, according to Mr. Butler, identified tracts he was interested in and "put up the money, but they [associates] were to prove up on it and then sell it back to him at a certain price."[60]

The contiguous San Gabriel Valley properties plus Rancho La Cienega constituted the expansive Baldwin Ranch. Lucky Baldwin, commented an interviewer, "can drive eighteen miles in a straight line across his acres."[61] Today, the San

Gabriel Valley cities of Arcadia, Monrovia, Sierra Madre, Temple City, much of El Monte and South El Monte, the City of Industry, most of Baldwin Park, Bassett (south of Valley Boulevard), West Covina and La Puente (west of Glendora Avenue), and portions of Montebello, South San Gabriel, and Monterey Park occupy former Baldwin lands, while from the La Cienega acreage there emerged the Baldwin Hills, Angeles Mesa, and Leimert Park developments. With his Comstock fortune, Lucky Baldwin "invested all but a million in Los Angeles County real estate and the Baldwin Hotel," according to Hiram Unruh, E.J.'s Southern California manager.[62]

THE BALDWIN HOTEL AND THEATRE—POWELL & MARKET STREETS.

Chapter Two

The Boom of the Eighties

The decade of the 1870s was a period of intense investment for Elias Baldwin, and while money spent in anticipation of future gain was not permanently forfeited, the drain on the millionaire's available resources became a topic of concern. Baldwin's investments, both business and personal, in Northern as well as Southern California, were on a grand scale. As Southland developments continued apace, E.J., in 1877, completed his magnificent Baldwin Hotel in San Francisco at a cost of $3 million.[1] A monument in which he placed great pride, the Baldwin Hotel was nevertheless a financial liability for Lucky in post-Comstock San Francisco, and it continued so for most of its years of operation.[2]

Jennie Dexter Baldwin

In 1879, E.J. escorted a new wife, Jane Virginia (Jennie) Dexter, and their three-year-old daughter, Anita (named for the Santa Anita Ranch), into a family suite at the Baldwin Hotel. Lucky had separated from his second wife in1875, no doubt in anticipation of the birth of Anita, January 10, 1876.[3] Extrication from his second marriage cost E.J. a $1 million settlement plus his Geary Street home in San Francisco,[4] but it won for him the girl that most associates agreed was "the one great love among his many purported loves."[5] E.J., already made a grandfather by his daughter, Clara, married twenty-one-year-old Jennie Dexter in 1879.[6] Their marriage seemed a happy one, though, sadly, they had only a short time together. On November 16, 1881, the fragile Jennie Baldwin died at the Baldwin Hotel. Throughout his life, Lucky kept a pair of her shoes and a pair of her gloves in memory, "treasured in a cabinet that attained almost the sanctity of a shrine in his Santa Anita home."[7]

The loss of his young wife was not the only blow E.J. endured as the bloom of his millions began to fade. It was shortly after Lucky had invested "all but a million" of his Comstock fortune in Southland real estate and in the Baldwin Hotel that in 1878, according to Hiram Unruh, "an unfortunate venture in stocks cost Baldwin $1.5 million, which took his million in the bank and made necessary a loan for which the hotel and Los Angeles county properties were mortgaged."[8] Raising the mortgage payment was soon

Anita Baldwin, Age 4

Hiram A. Unruh

to become an integral part of Lucky Baldwin's financial life. So, too, was Hiram Unruh.

Hiram Augustus Unruh (1845-1916), "the man whose name was inseparably linked with the romantic story of 'Lucky' Baldwin's rise to riches,"[9] was the nephew of Sarah Ann Unruh, E.J.'s first wife. Trained as a telegrapher and railroad agent, Unruh was plying his trade when he "was discovered by Lucky Baldwin."[10] "When I entered his service in 1879, he was $750,000 in debt," remembered Unruh. "He continued to borrow and gained the reputation of being a hopeless borrower . . . surrounded by grafters and leeches. I had to stand continually between him and the raiders."[11] H.A. Unruh was hired to help unravel the financial difficulties of the Baldwin Hotel, and he performed so admirably that E.J. came to rely with increasing regularity upon the younger man's sagacity and financial pragmatism. Lucky Baldwin "never worried," commented the *Los Angeles Times*. "He let Mr. Unruh do that."[12]

The Baldwin Hotel was leased to a private party in 1881, and Hiram Unruh was dispatched to Southern California to serve as Baldwin's "general business agent,"[13] a position he occupied not only through Baldwin's lifetime, but as executor of the estate as well. Lucky Baldwin needed cash in 1881, and Southern California land was a salable commodity. From the day of his first Southland purchase, in fact, E.J. had planned to subdivide and sell unused acreage. Five days after his Santa Anita purchase, the *Express* had reported that Baldwin would "cut up some five thousand acres into lots,"[14] and a few months later the *Herald* printed rumors that not only would Baldwin "divide his ranch into forty and eighty acre tracts, and build a house on each subdivision," but that he was arranging as well "for a narrow gauge railroad [to run] from this city to Santa Anita ranch."[15]

The railroad did not arrive, nor were the subdivision tracts then laid out, but in February, 1881, Baldwin did indeed sell some of his prized Santa Anita land. Nathaniel Carter, a pioneer Southern California boomer, touted in Los Angeles newspapers as "the famous excursionist,"[16] was the land buyer. E.J. chose well, the sale of an attractive piece of land to a booster of Carter's consummate skill boding well for the future of property values throughout the San Gabriel Valley. Mr. Carter purchased 845 acres of what became the Sierra Madre Tract from Lucky Baldwin for $33,881, and the deed of April 25, 1881, conveyed, as well, one half of all water from Little Santa Anita Canyon (with the water to be piped from Richardson's flume to Lot 23 of the Sierra Madre Tract, and there apportioned between the two properties).[17] Carter immediately divided his land purchase into small acreage tracts and announced terms of sale at one-third down, the balance due in one or two years at eight percent annual intereet.[18] By 1882, twelve families had settled in Sierra Madre, and the town had opened its first hotel.[19]

By 1883, a second piece of property joined Sierra Madre on the market, and at the same terms of sale. E.J. Baldwin's three thousand acre Santa Anita Tract was surveyed to the east of Sierra Madre in 1882-83 and was opened for sale with boomer Carter, interestingly, acting as agent for Baldwin.[20] Harking to E.J.'s Santa Anita Ranch as "an example of what work and good taste can do in the way of making a home in this delightful climate,"[21] Carter advertised not only the fertility of Santa Anita soil, but the natural beauty as well. Many of the "large white oak and live oak trees," he promised nature seekers, "will be retained along the avenues for shade and ornament." [22] For the more practical-minded, the advertisements informed all land seekers that "water is brought along the avenues to each lot in pipes from the Santa Anita and Sawpit Canyons, and each purchaser will own a share of the streams."[23]

Baldwin had acquired rights to Santa Anita Canyon water with his $200,000 land purchase from Newmark in 1875, and half the waters of Sawpit Canyon were his as well, a legacy of Louis Wolfskill,[24] who had appropriated the canyon water in 1869 in partnership with Alexander Weill of the Duarte Rancho.[25] Weill's successor was L.L. Bradbury, who, jointly with Baldwin, circa 1884, contracted to improve Wolfskill's dirt ditch irrigation system with more efficient tile pipe.[26] For buyers of the thirty and forty acre lots of the Santa Anita Tract ("which could be subdivided to suit"), it was guaranteed that water from Baldwin's half of Sawpit Canyon plus half the water from Santa Anita Canyon would be "deeded with the land in proportion to acreage, one share to each acre."[27] This arrangement was formalized July 10, 1886, with the incorporation of the Santa Anita Water Company as a regulatory body, with further provision forbidding the sale of water shares except as an "appurtenant to the land."[28]

Nathaniel Carter's first major sale of

Baldwin's Santa Anita Tract land came in April, 1884, and the buyer, a railroad builder in search of a country home, was William Monroe.[29] Wooed by both Carter and Baldwin, Monroe agreed to pay $15,000 for four thirty-acre lots located on either side of White Oak Avenue [Foothill Boulevard] and near the eastern ranch boundary.[30] A year later in March, 1885, apparently pleased with country life, Monroe added ninety acres more, and his brother, C.O. Monroe, purchased thirty acres of the Santa Anita Tract at $125 an acre.[31] The birth of Monrovia ("via Monroe's"), destined to become "perhaps the most successful small boom town of the San Gabriel Valley,"[32] was still a year in the future in 1885. Its meteoric rise, like that of the other sixty-odd boom towns put on the Los Angeles County market in 1886-87,[33] awaited the completion of the all-important railroad link.

On November 24, 1885, Elias Baldwin executed a contract with the Los Angeles and San Gabriel Valley Railroad Company, then building east from Pasadena, to extend their line across both ranch and Santa Anita Tract lands.[34] Organized August 30, 1883, "for the purpose of constructing, operating, and maintaining an independent broadgauge railroad from Los Angeles through the San Gabriel Valley to Mud Springs [San Dimas],"[35] the Los Angeles and San Gabriel Valley Railroad was completed to Pasadena on September 16, 1885, and extended to Lamanda Park on November 7 of the same year.[36] The road, it was projected, would both furnish residents of the valley with "a quick and easy access to Los Angeles" and "enable the through shipment of freight from any point in the valley to San Francisco without breaking bulk, at the least possible rates."[37]

The potential was alluring, and once convinced of the line's credibility, Baldwin was prepared to make a cash investment in its future. By terms of his contract, E.J. granted to the Los

Passenger Train Between Pasadena and Los Angeles, 1885
Los Angeles and San Gabriel Valley Railroad

Angeles and San Gabriel Valley Railroad Company, and to "their successors and assigns," without condition, the right to survey a line across his Santa Anita acreage. He further agreed to pay to the railroad "the sum of ten thousand dollars ($10,000) U.S. Gold Coin," to convey a right-of-way fifty feet wide across the property, and to deed two separate tracts of land "for depot, station houses, side tracks, and platform," under the express condition that the railroad "be completed and said trains running thereon within seven months" (unavoidable delays excepted) of the execution of the contract.[38] The two depots, one to be located "on Baldwin Avenue" (the main ranch thoroughfare) and the other "on Santa Anita Tract," were to be constructed by the railroad company, which would henceforward "ever cause local freight and passenger trains to stop regularly at both of said stations."[39]

The depot provisions were more than a convenience for Lucky Baldwin. By contract language, the Santa Anita Tract station was to be "within the boundaries of Falling Leaf Avenue [Huntington Drive], Santa Anita Avenue, and East Avenue [Mayflower Avenue, Monrovia], in "the townsite to be located by said Baldwin, on the line of the railroad."[40] The townsite was "Baldwin."

The first of the new year, 1886, brought daily advertisements to Los Angeles newspapers for the new "town of 'Baldwin,' now being surveyed on the Santa Anita Tract." [41] Surrounded by thirty-acre farm parcels, "Baldwin" was platted into "town lots and villa sites" in the spring, the smaller lots bounded by Falling Leaf Avenue and Diamond Street on the north and south, by Tenth and Fourth Avenues on the east and west.[42] Reflecting Baldwin's confidence in greater profits, prices were advertised as $175 to $200 per acre and "rapidly increasing in value."[43] E.J. also confidently noted in his advertisements that his namesake town was "soon to be traversed by the San Gabriel Valley Railroad [who] have arranged to locate their depot at 'Baldwin.'"[44]

The Los Angeles and San Gabriel Valley Railroad reached Santa Anita Avenue on June 1, 1886, and continued energetically on,[45] but neither the town nor the depot of "Baldwin" were to become part of its system. "Baldwin" died aborning, its demise perhaps indicated in the observations of a *Herald* reporter who accompanied the railroad graders and track layers on their push across the San Gabriel Valley. Surveying the town of "Baldwin," "located near the wash of Santa Anita creek," the reporter pronounced that though "the Santa Anita probably contains as much good land as any equal expanse of territory in Los Angeles county . . . the best portion has by no means been selected as the site for the town of Baldwin."[46]

"A short distance from the nascent city of Baldwin," he continued, "lies the young and vigorous settlement of Monrovia." [47] Though poor location must surely have contributed to the failure of the Baldwin townsite, a legitimate case could as readily be made that "Baldwin" was simply overshadowed by the nearby and dynamic Monrovia. In December of 1885, Lucky Baldwin sold 363 acres bordering East Avenue and his townsite to E.F. Spence, John Bicknell, and J. F. Crank.[48] The three influential men, reported the *Times*, "are making room for a settlement of 2500 people."[49] In the spring, Spence, Bicknell, and Crank joined William Monroe to lay out a sixty-acre townsite centered at Orange and Myrtle Avenues, and on May 17, 1886, the four men put the first town lots on sale there.[50] Monrovia grew rapidly. "Improvement has been great and constant," it was reported at the end of the year, when the town could boast a newspaper, three hotels, and two churches, as well as numerous small stores.[51] On December 8, 1887, Monrovia was incorporated as a city of the sixth class.[52]

Little wonder that "Baldwin" became but a memory. Baldwin, the man, likewise appeared to fade from direct participation in the real estate promotions of 1886, though it is doubtful that his withdrawal was more than a cosmetic, perhaps finance-related move. In February of 1886 Elias Baldwin deeded, for $350,000, all the remaining unsold portion of the Santa Anita Tract (approximately two thousand acres) to his manager Hiram Unruh.[53] It was Unruh who withdrew the "Baldwin" townsite advertisements in April, 1886, and it was also he who in September of 1886 had the two thousand remaining acres surveyed as "Part of the Santa Anita Tract."[54] One thousand of the acres, in five to twenty acre lots, were put on the market late in1886 at $250 to $500 per acre, and Unruh, rather ignominiously, advertised "Monrovia Station," a temporary stop located on the East Avenue boundary between Monrovia and the deceased "Baldwin," as the nearest point of departure for prospective purchasers.[55]

There were apparently few takers, but the new year brought a new spirit to Mr. Unruh's advertising. The survey map of "Part of Santa Anita Tract" had shown a large "X" drawn through a block of land to the northwest of "Baldwin." Located between Orange Avenue [Colorado Boulevard] on the north and California Street on the South, between Santa Anita Avenue on the west and Second Avenue on the east, and including several adjacent, though smaller "X'ed" lots,[56] the reserved acreage was to become the focus of a renewed and much more specific advertising campaign. Though the earliest advertisements did not put a name to it, Baldwin and Unruh's planned developments would become the boom townsite of Arcadia in April, 1887. In January, the pre-Arcadia advertisement appeared in local newspapers as the following:

SANTA ANITA TRACT

Part of the Celebrated Santa Anita Ranch, Now on Sale

200 ACRES, IN FIVE TO TWENTY-ACRE LOTS
AT $400 TO $600 PER ACRE, WITH WATER.

Mr. E.J. Baldwin has, at great expense, conducted an abundance of pure mountain water in pipes to the above tract, on which MONROVIA is located, and under the same water supply, vis: The Santa Anita Water Company's system.

Broad Avenues Already Open and Graded!

Many of them lined with Eucalyptus and Pepper trees. The rapid improvement of that portion now sold is an evidence of the estimation in which it is held. No expense will be spared to make the locality attractive and satisfactory to settlers, it being the intention of Mr. Baldwin to run

A Motor Railroad Along the Entire Length of Santa Anita Avenue.

Schoolhouse, Church and Public Park grounds will be set aside and improved. The contract with the San Gabriel Valley Railroad Company calls for the erection of substantial depot buildings.

An Air Line Railroad Will Soon Be Completed

To Los Angeles and trains run so that the tract will be only 30 minutes' ride from Los Angeles, the time via San Gabriel Valley Railroad being at present about 50 minutes.

See the tract before buying elsewhere. The Land and Locality speak for themselves. No Frost, No Fogs, No Dry Winds. Climate Perfect. Terms reasonable.For further information apply to

H. A. UNRUH
57 No. Spring St. Los Angeles, California[57]

The Los Angeles and San Gabriel Valley Railroad contract was an important one, and was shortly to become even more so for E.J. and Hiram Unruh. By January, 1887, the road was completed from Los Angeles twenty miles eastward to Duarte. Five passenger trains traversed it daily, making nineteen stops in twenty miles, and serving a Valley population that had increased by three thousand in the previous year alone. [58] The *Times* reported the line's earnings as "upward of $15,000 per year per mile," third in the country when measured in money earned in proportion to costs.[59] The figures were to increase significantly. On May 20, 1887, the Los Angeles and San Gabriel Valley Railroad became part of the transcontinental main line of The Atchison, Topeka, and Santa Fe, consolidated, along with seven other local lines, into Santa Fe's California Central Railway Company.[60]

The boom days of the late 1880s were sparked by the appearance, or more precisely, the competitiveness of the Santa Fe Railway Company. Passenger rates to Southern California plummeted as Santa Fe and the Southern Pacific vied for traffic. In 1887, Southern Pacific brought 120,000 people to Los Angeles; Santa Fe unloaded three to four passenger trains daily.[61] Nothing was impossible in a time when $25 or less[62] could buy transportation from St. Louis to Southern California, and railroad rumors were rife. "The Atchison, Topeka, and Santa Fe will construct an air line to Los Angeles," reported the *Monrovia Planet*,[63] and Unruh's Santa Anita Tract advertisement echoed the promise. The air line was surveyed just north of the Los Angeles and San Gabriel Valley road and was intended to cut off the line's sharp bend south out of Pasadena by constructing a road directly from Monrovia to Raymond Station.[64] The effect on both Monrovia and Unruh's proposed settlement was negligible, however, for the cut-off was eventually engineered to the west, at Lamanda Park, and shortly thereafter was eliminated altogether in response to public outrage in Pasadena.[65]

Equally without effect were the persistent rumors of a Southern Pacific branch line to be constructed along the foothills north of the Santa Fe air line.[66] Rumor and hyperbole were integral parts of boom psychology, however, and Unruh and Baldwin played the game with ease. In late March, "Arcadia" was chosen as the name for their already ballyhooed Santa Anita Tract townsite. A "euphonious" appelation in the estimation of the *Los Angeles Herald*,[67] the selection of "Arcadia" probably had little more significance than simple euphony and perhaps a low-key association with the pastoral simplicity of Arcadian Greece. Interestingly, however, "Arcadia" may not have been the first choice of name for the townsite. In January of 1887 the *Herald*, reporting on the Southern Pacific branch survey, noted that the line would run "through Pasadena, Lamanda Park, Live Oak, Monrovia, Duarte, Azusa, and San Dimas." [68] In corroboration, and just two weeks prior to the "Arcadia" christening, the *Monrovia Planet* informed its readers that "the new townsite of 'Live Oak' has been laid out on Santa Anita Avenue."[69] The reasons for the rejection of "Live Oak," if indeed it was ever seriously considered by Unruh and Baldwin, are as obscure as the origins of "Arcadia." It was without explanation that on March 26, 1887, the *Planet* quietly announced, "E.J. Baldwin's new town is to be named Arcadia."[70]

Arcadia "is one of the most delightful spots imaginable for a town," enthused the *Los Angeles Herald*. "At present [March 30] the city consists of a sign under a huge spreading oak tree, but in four days about three hundred lots have been sold and the surveyors are having hard work to keep up with the agent, Mr. H.A. Unruh."[71] For buyers who remembered the aborted "Baldwin" townsite, advertisements guaranteed "No wash, no rocks!"[72] "Pressure water piped to every lot" was further inducement to those who flinched at the steep $400 to $600 per acre sales prices. [73] Not everyone flinched. On April 9, the *Monrovia Planet* announced that "nearly six hundred town lots have been sold. There are none left above Falling Leaf Avenue."[74]

By April 23, it was reported that "about four-fifths of Baldwin's Arcadia townsite has already been sold, and prices for the remaining property have been advanced."[75]

"Mr. Baldwin announces that all money he receives from sales of lots will go toward improving the town."[76] It was reported that E.J. was personally supervising the work at Arcadia,[77] and from the scope and occasional flamboyance of proposed developments, it seems likely that it was indeed he not only orchestrating, but also financing the birth of Arcadia. Town necessities were not neglected. A wood depot, located where the California Central (Santa Fe) crossed First Avenue, was completed by June, 1887,[78] in apparent fulfillment of E.J.'s Los Angeles and San Gabriel Valley Railroad contract, and several houses, including that of Hiram Unruh at First and Falling Leaf Avenues, were constructed with local lumber and brick. Baldwin's Santa Anita brickyard at the ranch supplied over two million bricks for Arcadia buildings.[79]

Arcadia's Southern California Floral and Perfumery Company, its headquarters located at Santa Anita and Orange Avenues, reflected Lucky Baldwin's interest in horticulture. Baldwin and Hiram Unruh were reported as "among those interested in the project"[80] when the Floral Company purchased thirty acres in the Santa Anita Tract, and with incorporation in March of 1887, H.A. Unruh was installed as company treasurer.[81] President Emory Smith, an experienced nurseryman whose intention it was "to make this one of the largest 'flower gardens' in America," and Dr. W.H. Masser, the secretary-chemist in charge of manufacturing perfume from the flowers, completed the governing board.[82]

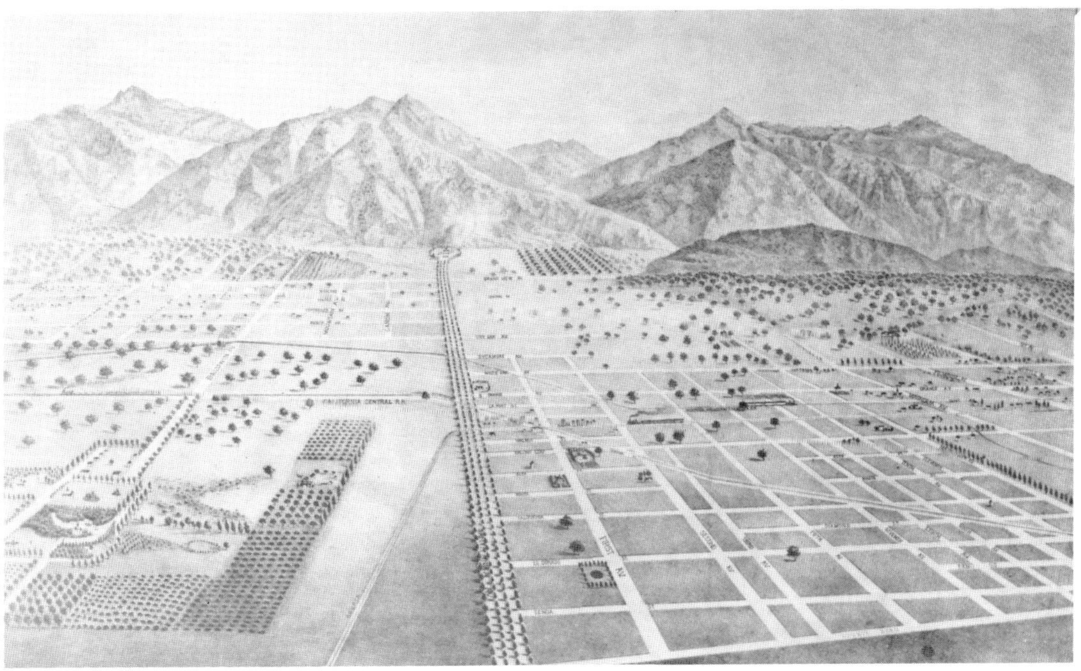

Birdseye View of Arcadia and Santa Anita Tract

The Floral Company imported rare trees and plants to add to local stock offered for sale, and on their eleven acre Santa Anita and Orange Avenue lot (their remaining acreage was foothill land), they surrounded the company headquarters with extensive public exhibition grounds. "Everything will be very attractive and the place will no doubt become a resort for lovers of flowers," summarized a Monrovia reporter.[83]

Lucky Baldwin's love of nature found additional outlet in his active campaign of tree planting along townsite streets. An estimated forty thousand pepper and Eucalyptus (blue gum) trees had been planted along Santa Anita Tract avenues in anticipation of the sales of 1886, [84] and with the genesis of Arcadia, Santa Anita Avenue became a focus of attention. Its six-mile length, lined with a double row of Eucalyptus, was to be extended an additional six miles to connect with the Southern Pacific depot in the valley, and a "motor road," it was reported, would run the entire distance.[85]

The "Santa Anita Avenue Railway,"[86] according to projections, would terminate at the mouth of Santa Anita Canyon, where a "monster hotel" was contemplated.[87] "Large enough to place the Raymond inside of it," Hotel Santa Anita was to be the showpiece of Arcadia, a testament to Baldwin's grand design.[88] The building was to be four stories high, with a five hundred foot frontage, constructed of canyon granite, and complete with pleasure trails and roadways for its guests.[89] "Baldwin estimates its cost at $500,000 and says it will be the most complete hotel in Southern California," said the *Planet*.[90]

Hotel Santa Anita was never built. The cost, no doubt, was prohibitive in E.J.'s weakened financial position, and he turned instead to a set of plans for a lesser, though still quite substantial hotel in the heart of his Arcadia townsite.

Arcadia Depot

Hotel Oakwood was begun in June, 1887 on the corner of First Avenue and Santa Clara Street, across the tracks from the Arcadia Depot.[91]

Contractor Frederick Sandham of Pasadena was engaged to construct the two-story brick building,[92] and when it was completed in the latter part of the year each of its thirty-five rooms (five hundred rooms had been planned for Hotel Santa Anita) could boast "hot and cold water, a fireplace, and all the accommodations."[93] By 1888 there were telegraph and telephone connections, an express office, and a post office off the lobby, as well as an acclaimed bar and restaurant with food and wines "served fresh from the ranch."[94]

To connect their fledgling Arcadia with an already well developed business district in neighboring Monrovia, and to make more valuable Santa Anita Tract lands lying between the two towns, E.J. and Unruh joined William Monroe and E.F. Spence in May of 1887 to incorporate the local Arcadia and Monrovia Railway. The three-mile steam line was to run from First Avenue in Arcadia along Santa Clara Street and Monrovia's Chestnut Avenue,[95] but before actual construction was even begun, a more significant project was revealed.

On July 9, a meeting was held in the office of General W.A. Pile (a Monrovia resident) to discuss construction of a narrow gauge steam railway from Los Angeles, through Alhambra, San Gabriel, and Arcadia, to a terminal in Monrovia. Baldwin, Unruh, Arcadia, and Monrovia had subscribed $27,500 to the undertaking, according to General Pile, and both of the towns were already surveyed.[96] Reflecting their interest, William Monroe and Hiram Unruh were elected to the Board of Directors of the new Los Angeles and San Gabriel Valley Rapid Transit Railroad Company in July, 1887.[97]

Hotel Oakwood

The Rapid Transit was begun in Monrovia (where the surveyed Arcadia and Monrovia Railroad was merged with the larger line) and within a year it reached the eastern boundary of Los Angeles.[98] The formal opening of the "Jack Rabbit Line," as it was popularly known, took place in August, 1888, with passengers transferring to free carriages at Mission Road to complete their trip to the city. The Rapid Transit was never able to secure a right-of-way across the Los Angeles River.[99] The Los Angeles and San Gabriel Valley Rapid Transit Railroad Company was taken over by Southern Pacific in 1893, but continued to serve Valley residents with four round trips to Los Angeles daily, and at faster times than possible on the Santa Fe.[100]

By the end of 1888, Lucky Baldwin's Arcadia could boast of two railroad connections, each with frequent service (Santa Fe ran five through passenger trains daily),[101] of its fine Hotel Oakwood, of the Southern California Floral Farm, and perhaps a few homes, but, in truth, very little else. In his application to establish a post office in Arcadia in 1888, Daniel Requa numbered "about 150 inhabitants" in the town.[102]

Both Monrovia and Sierra Madre had active newspapers, churches, schools, and social groups. Baldwin's town had none. Arcadia was young, however, and oriented more toward a tourist and small farming economy than to the small businesses and city-living that characterized Monrovia. Arcadia in 1888 was perhaps more functional as a selling point for E.J.'s growing number of subdivisions than as an organized townsite.

With no motive beyond money-raising, Baldwin had subdivided and offered for sale fifteen hundred acres of his Felipe Lugo lands in 1886 in a joint undertaking with Richard Garvey,[103] and in 1891 he proposed a similar money-raising subdivision of lands south of the Arcadia townsite. "The property of E.J. Baldwin," advertised Mr. Unruh, "is being laid out and subdivided as fast as required and, where needed, supplied with water suitable for fruit and other farming."[104] In addition to remaining Santa Anita Tract lands, acreage in "Santa Anita Colony," "Additions No. 1 and No. 2 to Santa Anita Colony," and "Parts of San Francisquito, Felipe Lugo, and Merced Ranchos" were put on the market by Baldwin. "Other lands in tracts of from five hundred to ten thousand acres," it was noted, were available as well "on special terms."[105]

Located between Duarte County Road (Duarte Road) and Azusa County Road (Lower Azusa Road) and bounded by El Monte Avenue on the west and Second Avenue on the east,[106] the eight hundred acres of "Santa Anita Colony" were offered to buyers in 1891 at two hundred dollars per acre—25 percent cash, balance due in six years at 7.5 percent annual interest—with the provision that the land be planted to citrus within a year.[107] Stock trees, buyers were informed, were available at market prices, on the "same time, payment, and interest as for the land," from the Southern California Floral Company in Arcadia.[108] The "Additions to Santa Anita Colony," located between El Monte and Baldwin Avenues,[109] put another seven hundred acres on the market later in the year. The unimproved acreage was offered at $150 to $200 dollars per acre, and without the planting requirement, while in the San Francisquito, Felipe Lugo, and La Merced Ranchos three thousand miscellaneous acres were put on sale under identical terms.[110]

"The present is a good time to buy land," the advertisements read, "as the bottom has been reached."[111] In blunter terms, the boom of the eighties was over, and with its demise went the inflated land values upon which it had been predicated. Townsites languished, or simply disappeared in the wake of the boom, but as the Baldwin land advertisements indicated, the productivity of the soil was not correspondingly diminished.

E.J. had an abiding faith in the efficacy of land investment, and his advertising of 1891-92 continued to be unreservedly optimistic. According to Baldwin, "California's Choicest Locality" was the San Gabriel Valley, and anyone sampling his brochure's comments could confirm the fact.

> An average of more than three hundred sunny days makes it possible to work in the open air the greater part of the year . . . epidemic diseases, poisonous insects, tornadoes, cyclones, earthquakes, and thunderstorms are practically unknown . . . all productions of Eastern and Northern states can be grown here, besides those of a semi-tropical and many of the tropical countries. . . . a comfortable house, hard finished, of four to six rooms, can be built for $600 to $700. . . . twenty acres of bottom land will support an average family better, and with less labor, than 160 acres of farming land in the West and Northwest.[112]

Eucalyptus-lined Road

Chapter Three

The Ranch—Produce, Pride, and Profits

The pride and enthusiasm evidenced in the Baldwin land advertisements were nowhere better displayed than at E.J.'s Santa Anita home ranch. The four thousand remaining acres of Rancho Santa Anita were in "a high state of cultivation" in the year 1891,[1] and Baldwin skillfully utilized not only the productive potential, but also the attendant publicity afforded his acreage. Santa Anita was a tourist attraction and the focal point of the Baldwin Ranch, and though the productions of E.J.'s other lands were included in crop and stock totals, Santa Anita held the unrivalled lead in output as well as beauty. It was to the "Home Place" that writers and observers invariably turned their interest.

An early observer had written of the ranch that it was "the intention of the proprietor to make [Santa Anita] one of the most complete and romantic spots in the country."[2] Careful thought and a great deal of money were indeed expended on the ranch, and sixteen years after its inception, H.H. Bancroft not only cited statistical proof of its productivity, but gave additional, rhapsodic testimony to the success of Lucky Baldwin's corollary campaign of beautification.

> The scene is one of fairy-like loveliness; not only the little bijou residence and its surroundings, but the entire estate, with its groves and vineyards, its golden fruit and waving harvests, its shaded drives and vistas of mountain peak and valley, carrying the beholder into an ideal region, calm and peaceful as the fabled realm of Rasselas, where soft vernal airs induce forgetfulness of the din and turmoil, the crowded streets and selfish intensity of city life.[3]

Mr. Bancroft, as was his practice, did not personally appear upon the scene of "fairy-like loveliness." An otherwise unidentified "Mr. Fraser" was the apparent Bancroft emissary to Santa Anita in 1891, and it is his recorded, first-hand observations, plus information garnered from and by others, that provide firmer evidence of a ranch of inordinate pride, respected produce, and steady profit. A visual reconstruction of Baldwin's Santa Anita Ranch, drawn from memory by nonagenarian, Elizabeth Cleminson, whose family came to the ranch in 1889,[4] adds a further dimension to the verbal and statistical descriptions, and it is through the combined media that the clearest representation of the ranch emerges.

Mr. Fraser "alighted at Santa Anita Depot" on August 26, 1891, and boarding a carriage, "started off to visit as many of the places of interest on the ranch that limited time would allow."[5] He might well have started his investigation with the recently completed Santa Anita Depot itself. Located on Baldwin Avenue, and one of the two freight and passenger structures promised Baldwin in his Los Angeles and San Gabriel Valley Railroad contract, the ranch station seen by Mr. Fraser was not completed until November of 1890, three years after E.J.'s Arcadia Depot on First Avenue. The reasons for the delay of the Santa Anita Depot are complex and not fully detailed, but the apparent cause was a prolonged discussion among Baldwin, Santa Fe, and peripherally, Sierra Madre, over the design

Map of Baldwin Ranch

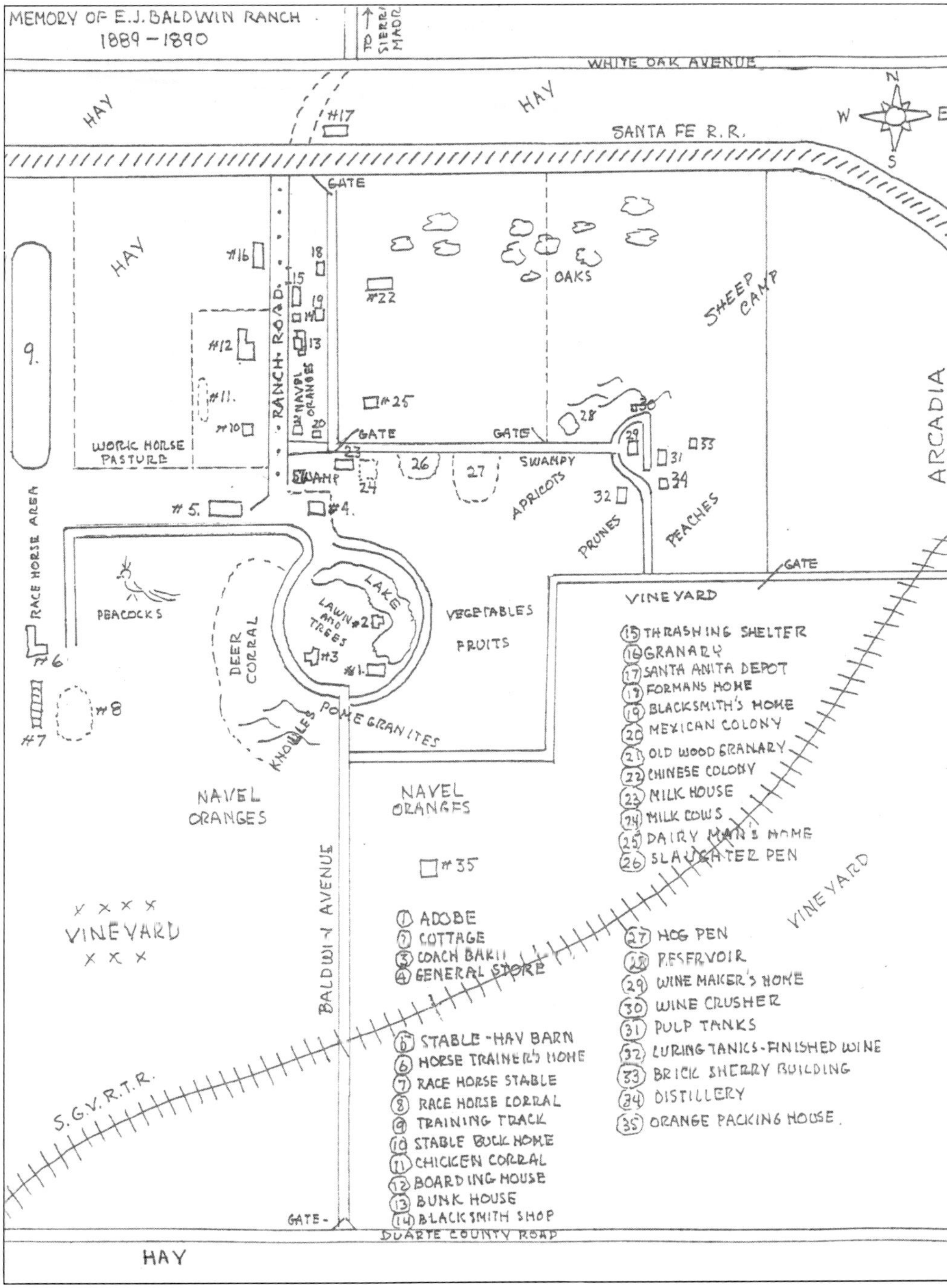

and cost of the proposed structure.

In June, 1886, the Los Angeles and San Gabriel Valley Railroad track was completed to Santa Anita, and a temporary station (a covered bench) was erected to serve both ranch and Sierra Madre Tract residents.[6] A sidetrack was laid from the station to Baldwin's brickyard nearby in the spring of 1888,[7] and at that time plans for a permanent depot building were drawn.

Mr. Unruh showed the plans to Sierra Madre town fathers in April of 1888 and, according to the *Sierra Madre Vista*, apprised them of an estimated $5000 cost for the structure, "of which," the newspaper said, "Mr. Baldwin agrees to pay $2500, the railroad $2000, while we are asked to contribute $500."[8] A year later there was not yet a permanent depot, and L.E. Maidhof, of Sierra Madre, felt the need to explain the impasse in the *Vista* of March 9, 1889. Though "the railroad company is quite willing to erect an ordinary building," he said, "Baldwin will give no title [to the ground] except with the condition that the railroad company erect an elaborate building . . . and with the further proviso that the station is to be called Santa Anita."[9] The cost split was to remain the same, but according to Mr. Maidhof, it was the depot name against which Sierra Madrans rebelled. "They naturally object to paying money toward that which will not advertise them," he said.[10]

That the controversy over name was resolved in favor of "Santa Anita Depot"[11] is certain. What is less certain is the resolution of costs and origin of design of the station. Local newspapers were silent until August of 1890 when the *Sierra Madre Vista* reported "the Santa Fe railroad depot for Santa Anita is an approaching reality." [12]

Santa Anita Depot with Signal Arm

An "elaborate" two-story depot, said the *Vista*, was to be constructed at Baldwin Avenue using one hundred thousand Baldwin bricks[13] which were later reported as "donated and placed on the grounds" under E.J.'s orders.[14]

A Mr. Logie, "who built the new Redondo depot," was mentioned as the builder,[15] and one can surmise, though not with certainty, that the Santa Anita Depot was a Santa Fe design project that accepted donated Baldwin brick in payment for implementation of the more elaborate building plan E.J. had desired. Also possible, however, is a scenario in which Baldwin not only suggested the design carried out by Santa Fe, but furnished the brick, not in addition to, but rather in lieu of his $2500 contribution.

Whatever its genesis, the Santa Anita Depot opened for business on November 23, 1890, with Mr. M.C. Taylor, newly transferred from the Arcadia Depot, as agent and first resident of the second-story depot living quarters.[16] Among the innovations at the station, according to the *Vista*, was "a patent lever train signal, which is easily worked by the agent . . . to flag all trains, when necessary, including the Overland."[17] Baldwin, explained the *Vista*, played a personal role in arranging to have Santa Anita "made a flag station for the Overland passenger,"[18] and, in corroboration, Santa Fe timetables published in the local newspaper denoted a daily through train that would "stop on signal" for passengers at Baldwin's Santa Anita station.[19]

As a one-day visitor, Mr. Fraser had little time to muse upon the historical background of his railroad stop. His eyes were turned south from the depot instead, down the main ranch road that he reported was fully forty feet wide and lined with Eucalyptus and pepper trees.[20]

Lucky Baldwin's ranch store was a quarter mile down the picturesque avenue, "a typical California general merchandize store," according to Mr. Fraser.[21] Advertisements for the Santa Anita store ran regularly in the *Sierra Madre Vista* and brought "a large trade from the surrounding country,"[22] as well as from ranch employees. "Dry goods and groceries, fancy goods, notions, woodenware, crockery, glassware, boots and shoes, carpets . . . [Baldwin] wines and brandies, and superior building bricks" were among items advertised and offered for sale.[23]

General Store

"Every person on the ranch had a book at the store," remembered Julian Fisher, "and on pay day, Baldwin would pay everybody down there."[24] Mr. Fisher's father, John Isaac Wesley Fisher, was Baldwin's blacksmith, one of several black employees on the ranch. "E.J. Baldwin has sixty Negroes en route from North Carolina to work on his great ranchos," the *Los Angeles Herald* reported in 1886[25] and, according to Mr. Fisher, several blacks still worked at Santa Anita at the turn of the century. The North Carolinians were employeed most frequently in the Baldwin stables as trainers, jockeys, exercisers, coachmen, and laborers, and housing for them was provided in the same area.[26]

John Fisher and Son Julian

Blacks were not the only ranch employees. Baldwin "had a kind of melting pot," commented Julian Fisher.[27] White laborers, usually craftsmen and managers, were a second group employed on the ranch, and typical was Samuel Butler, a single man given accommodation at the boarding house as he worked on a part-time basis boring wells for Baldwin.[28] Mexican employees, their ranks swelled at harvest times, lived as a group north of the store and worked most frequently as field hands.[29] Chinese workers, the final group in the "melting pot," worked in the orchards, at the winery, and as domestic servants on the ranch, and they, too, lived in a separate area, north of the dairy. "Probably one third of the labor now on the place is Mongolian," said Baldwin in 1891.[30]

The Baldwin work force labored six days a week, sunup to sundown,[31] the wide mix of their skills a reflection of the variety of ranch output. "'What do you raise on the ranch?' was a frequent question. 'Everything in the world but the mortgage,' was Baldwin's smiling answer."[32] Turning west into a crossroad from the ranch store enroute to his next stop, Mr. Fraser traversed a part of the Baldwin wheat, hay, and grain fields. "There are twenty-five hundred tons of hay raised on the ranch," Mr. Fraser was told, "of which we feed eight hundred tons to our own stock and sell seventeen hundred tons."[33] Alfalfa yielded fifteen to twenty tons per acre at Santa Anita, nearly two thirds of it used for home consumption, the rest sold. Wheat and barley were grown just south of Sierra Madre, and some twenty-eight thousand sacks of grain were annually stored in a nearby granary.[34]

The yields not sold were used to help feed ranch livestock, including the work horses and mules housed in a large feed stable near the crossroad intersection. Mr. Fraser had not come to see work horses, however. It was rather to the thoroughbred stable complex that he next turned his particular interest. Located near the mile training track at the western ranch boundary, the racing stables were entered through a gate arched over by a sign reading "The Home of Grinstead."[35] Grinstead and Rutherford, two stallions, were Lucky Baldwin's first thoroughbred purchases, acquired in Kentucky in 1874 for $18,500. Two running mares plus six yearling fillies were added to the stable the following spring, and when in 1878 E.J. purchased several additional brood mares, Santa Anita became a breeding farm in its own right.[36] By the time of Mr. Fraser's visit, the Santa Anita stud comprised seven stallions, fifty-four brood mares, and two geldings.[37]

Lucky Baldwin was a pioneer in breeding and raising thoroughbreds in California instead of the favored Kentucky bluegrass. He was also among the first to successfully race his western-bred stock on eastern circuits. John Fisher, "considered the finest farrier in the country,"[38] shod Baldwin's horses, and black jockeys Isaac Murphy and Freddy Welch were hired to teach E.J.'s stable workers to ride. "We not only raised our own race horses," said John Fisher, "but we raised our own jockeys."[39] Si McClain, Pike Barnes, and Willie Van Buren, all Derby winners,

Plowing Field at Ranch

were graduates of the Santa Anita jockey school.[40]

"Lucky Baldwin manages his stables as a business enterprise and he makes them pay," commented one impressed reporter.[41] Racing under his black with red Maltese Cross colors, E.J. shipped twenty-five to thirty horses a year to the eastern courses and averaged $100,000 annually in winnings.[42] Credited in turf history as "the most successful racing man in the country, considering the size of his stud,"[43] Baldwin also "was one of the less than six or so owners who actually made racing pay."[44] The mare, Los Angeles, purchased by Baldwin as a yearling for $3500, won over $100,000 for her owner before an accident ended her five-year racing career.[45]

Two "Pullman palace horse cars," the "Santa Anita" and the "Arcadia," fitted inside with comfortable stalls, transported the Baldwin racers to their scenes of triumph.[46] In 1885, E.J.'s stable "blanketed the nation," racing in St. Louis, Chicago, Saratoga, Sheepshead Bay, Jerome Park, Latonia, Louisville, Baltimore, Monmouth Park, and Washington, D.C., as well as in California.[47]

Baldwin horses won fourteen stakes in seven different states in 1885, and at Saratoga won fifteen of twenty-five starts.[48] Volante won the first of Baldwin's American Derbies in 1885. At the time, the American held more prestige than the Kentucky Derby. In 1886 and 1888, E.J. won again with Silver Cloud and with Emperor of Norfolk.[49] The Emperor, Lucky's favorite horse, raced only the one year of 1888, but won twenty-one of twenty-nine starts, earned $79,000,[50] and became the only horse ever to win all three of Washington Park's great three-year-old races (the American Derby, Sheridan Stakes, and Drexel Stakes).[51] In 1894, Rey el Santa Anita won Baldwin's fourth American Derby to set an owner's record that was never equalled.[52]

More prosaic than his racing stock, but

Home of Grinstead

Santa Anita Stud Barn

equally well regarded, was Lucky Baldwin's dairy stock of Durham, Jersey, Holstein, and imported Devon breeds. The Santa Anita dairy, "run entirely by steampower, and fitted with all the latest and most approved machinery,"[53] turned out two hundred pounds of butter a week, enough to supply both the ranch and E.J.'s hotels, with the remainder finding a ready local market.[54]

Baldwin kept large herds of cattle, sheep, and hogs, some at Santa Anita and the rest on his outlying ranches. The ranch slaughtered all its own meat, according to Mr. Fraser, and retailed the surplus to the public at ten to twelve cents a pound.[55] The Santa Anita poultry yard also served both ranch and retail purposes, with chickens selling at five to nine dollars a dozen, and eggs averaging twenty-five cents a dozen.[56]

Cash crops at Santa Anita did consistently well. In a Chamber of Commerce report for the 1891 season, Hiram Unruh indicated that even without fertilization, a single seventy-five acre orchard at the ranch had yielded eighty thousand boxes of oranges and netted a $102,895 return for farmer Baldwin.[57] Mr. Fraser could well believe the figures as, returning from his visit to the thoroughbred stables, he reported "interminable stretches of orange groves . . . and orchards upon orchards of different fruits" in the southern and western portions of Santa Anita.[58] There were five hundred acres of oranges alone at Santa Anita in 1891 and over one million young orange trees in foothill nurseries.[59]

In addition to oranges, moreover, almost every other type of semi-tropical and deciduous fruit was grown. "E.J. Baldwin's Santa Anita Rancho has 16,000 lemon and orange trees," reported the *Sierra Madre Vista*, "2000 pomegranites, 3000 English walnut trees, 5000 almond trees, 2500 peach trees, 4000 pear trees, 2000 apricot trees, 1000 fig trees."[60] To that list, Bancroft added Japanese persimmons, olives, and "camphor, pepper, coffee, and tea plants."[61] Baldwin did his own packing at Santa Anita and marketed his produce on his own label, "without," reported the *Times*, "being beholden to the fruit exchange or any of the commission men."[62]

The irrigation system for these Santa Anita

Maltese Cross

Orange Crate Label

productions, "planned by Mr. Baldwin without the assistance of engineers," according to Bancroft,[63] was dependent upon both local canyon and artesian waters. E.J. had retained half interests in Santa Anita and in Little Santa Anita Canyons during the boom of the eighties, and from the partitioners of each[64] he ran miles of concrete pipe to carry water to his property.[65] Further, some five hundred acres of Santa Anita lay in an artesian belt that crossed the ranch from west to east (the result of the dam-like effect of the Raymond Hill Fault). From both cienega streams and artesian wells on the acreage, Lucky collected additional water to distribute for irrigation.[66] A network of pipes and cement-lined ditches eventually criss-crossed the ranch, transporting cienega, well, and canyon waters first to small reservoirs, thence, as needed, to the productive Baldwin orchards and groves. "The Baldwin Ranch," opined a *Los Angeles Times* reporter, "is perhaps the best irrigated piece of property in the United States."[67]

Many observers found the Santa Anita Ranch among the most beautiful of properties as well. Mr. Baldwin "thinks there is nothing too good for Santa Anita," commented the *Sierra Madre Vista*,[68] and, in fact, there seemed to be little expense spared in either landscaping or building projects at the homesite.

Recalling her own pleasure in the beauties of the ranch, Mrs. Cleminson spoke of peafowl and guinea hens roaming free on manicured lawns, of Baldwin's deer corral, of the boulder-edged natural lake at the homesite, of rare plants and trees intermingling with the native oaks E.J. had insisted upon keeping on the land. "Even the

Irrigation Channel at Adobe

swamps were beautiful," she said, "with their cattails and willows, wild grapes and vines."[69] My father "fought to guard these great oaks that dot the fields of Santa Anita," said Anita Baldwin. "He loved trees. . . . He loved natural things."[70] And if you dared to cut a tree (not even for firewood!), kill a peacock, or mistreat an animal, added Julian Fisher, "you could be fired immediately."[71]

Mr. Fraser passed the "thirty odd occupants" of the deer park enroute to his next planned stop at the Baldwin homesite. Just past the deer park, and immediately to his left, Mr. Fraser found "hidden in a wall of evergreen hedges and surrounded by semi-tropical foliage," Mr. Baldwin's private coach barn in which, according to Fraser, E.J. "stored fourteen carriages of different styles for his private use."[72]

Parking his own conveyance, Mr. Fraser proceeded on foot toward the lake and E.J.'s lakeside homes. "A vast expanse of lawns laid out with brilliantly colored flower beds," he said, graced his path;[73] overhead the skyline was punctuated with beech, elms, maples, a grove of palms, a stand of Monterey pines, cypress, silk oak, English oak, acacias, and more.[74] A weeping willow overhanging the lake was said by Bancroft to have been cut from a tree at Napoleon's tomb.[75] Commented a *Los Angeles Times* reporter,

> I have seen a number of botanical gardens in different parts of the world, but there are few more beautiful than the grounds about this home of Lucky Baldwin. It is one of the prettiest places in the world, and every tree and shrub connected with it has been planted by his direction. He took this vast estate when it was practically a desert and he made it into a land of flowers, trees, and fruit-bearing orchards.[76]

An architectural critic might have been equally impressed with the fanciful Baldwin cottage.[77] "The new cottage being built for Mr. Baldwin is situated on a knoll next to a large lake," commented the *Pacific Rural Press* in the spring of 1885. An artesian well in front of the cottage, they added, "was throwing water six feet above the ground in a large volume, and cemented, open raceways were made for the overflow to pass off in little rivulets all around the house."[78] The cottage, "Baldwin's Belvedere," as the *Times* referred to it, was reported "ready for occupancy" in the summer of 1886.[79]

> On the bank of this lake has been erected a beautiful cottage of four large rooms and bath. From the main drive a handsome shaded arbor leads up to the cottage, where the visitor treads a marble walk to the marble steps which rise to a marble floored verandah. . . . furniture and bric-a-brac have been brought from the proprietor's rooms at the Baldwin Hotel in San Francisco. Marble mantels of Parian whiteness with deep set grates—in which it is intended to burn gas, made on the premises from

Coach Barn Ironwork

naptha—lend an air of solid comfort. . . . It is, however, in the windows that the aesthetic hand of the artist has found scope. . . . One of these contains a fine portrait of Byron, another of Shakespeare, a third a bathing scene. . . . Mr. Baldwin has also chosen to perpetuate the memory of the former Mrs. Baldwin, nee Dexter, not only by having an almost life size oil portrait of that lady, as she appeared in the halcyon days of maidenhood, hung in the bedroom, but in having a copy made on the glass that forms one of the doors. . . . The artisan in charge of the fitting up says he has performed similar services in Bonanza Flood's house at Menlo Park, where the windows were considered fine, but do not compare with Baldwin's though made in Europe, while the latter are of San Francisco manufacture.[80]

The Baldwin cottage, elegantly outfitted with bedrooms and sitting rooms, served as the ranch guest house. To cook there was unthinkable. "In order that the odor of the kitchen shall not invade the perfume-laden air," explained Mr. Fraser, "Mr. Baldwin has his dining hall and kitchen in a separate building."[81] Located across the lake from the cottage and much improved from the "little adobe hut" E.J. had found on his property in 1875,[82] an eight-room, L-shaped wood and adobe structure was not only the cookhouse, but also Baldwin's personal residence at Santa Anita. The "Adobe," as it was called, provided kitchen and servants' rooms in its old adobe wing, bedrooms, dining room, and parlor in its new wooden wing.[83] A rock arch led to the adjacent "Adobe" courtyard, where family and friends could be found engaged in a game of croquet, or perhaps enjoying a friendly glass of the Baldwin wines and brandies manufactured not far from the homesite.

The winery was Mr. Fraser's last stop in his Santa Anita tour, and a favorite with most ranch visitors. Raising a variety of grapes in his seven

Queen Anne Cottage With Mirror Lake

hundred acre vineyard, E.J. annually produced one hundred thousand gallons of wine, including angelica, port, zinfandel, claret, white wines and sherries, and thirty thousand gallons of brandy.[84] Orange champagne was a Baldwin specialty that became quite popular in San Francisco, where, in the words of H.H. Bancroft, it was "styled by connoisseurs the nectar of the gods."[85]

The quality of the Baldwin productions was hardly to be denied, as gold medals and blue ribbons accrued with every show and exhibition.[86] Baldwin "has made some of the finest exhibits individually of native products ever attempted by a Californian," wrote an experienced observer, "and not only all over the United States, but in several of our sister republics."[87]

The Baldwin winery at Santa Anita was comprised of one large brick building reserved solely for the storage of sherry, three wooden buildings for other wines and brandies, and in the rear of these, a distillery.[88] E.J. did all of his own coopering and bottling at the winery and, according to Bancroft, "absolutely refuses to sell his wines and brandies with a French label, as is the practice of many vintners."[89] Instead the trademark for Santa Anita Vineyards, molded on all wine bottles, was a Maltese Cross with the initials *EJB* proudly stamped in the quadrants.[90] "No visitor to California can afford to return home without placing his order for some of these rare old wines," admonished an advertising pamphlet.[91]

Mr. Fraser concluded his tour of the Santa Anita Ranch at the winery without mention of the outlying properties that constituted the bulk of Lucky Baldwin's landed estate. La Cienega, La Puente, Potrero Grande, Felipe Lugo, La Merced, Potrero Chico, and San Francisquito remained largely under-utilized by Baldwin in 1891. There were no townsites attempted on any of the properties, nor were there any apparent attempts at intensified agricultural development.

Portions of the lands had been offered for sale in 1891 and, in 1893, a *Times* reporter repeated gossip that a "Colorado syndicate" had offered, without success, $1 million for the

Baldwin Adobe, Adobe Wing

whole of the La Puente acreage.[92] Much of the unsold land, it seems, was rented, either outright or on shares.[93] In his advertising brochures, Hiram Unruh included a paragraph of rental possibilities that stated in detail that

> Lands can be rented for grain or hay at from $2 to $5 an acre; for vegetables and fruit, where water is supplied, at from $10 to $25. Land can also be rented on shares, generally from one-fourth to one-third of the crop.[94]

Lucky Baldwin built a warehouse on his land near the Southern Pacific depot at Puente to store share crops—hay in the dry years, barley and wheat when there was enough rain—paid by farmers who rented his lands in the area.[95] Several Vineland (today's Baldwin Park) farmers were renting adjacent Baldwin lands as well,[96] and in today's West Covina area, J.R. Elliott leased a reported five hundred acres for wheat farming.[97] Baldwin himself farmed parts of his outlying acreage. Mrs. Cleminson remembers barley land "between the San Gabriel River and La Puente,"[98] and in the El Monte/Felipe Lugo area, Baldwin had a hops ranch of note.[99] The Baldwin hay and wheat combines, Julian Fisher recalled, travelled beyond El Monte to reap the ranch harvests.[100]

Outlying lands, further, provided grazing for Baldwin livestock. The Temple and Workman sheep, augmented by Baldwin purchases (E.J. bought "a large quantity" of Southdown sheep in England in 1877),[101] populated parts of several ranches in the area north of today's Whittier. "Mr. Cameron was in charge of the sheep ranch for Baldwin," said Mrs. Cleminson, "and his home was right where the California Country Club is now, up on the knoll."[102]

South of El Monte, "when you got to Weaver and went into a swamp, and that clear on over to Rosemead Boulevard, was all fenced in," continued Mrs. Cleminson. "That was his big cattle ranch, where he raised his beef cattle."[103] South of Los Angeles, Rancho La Cienega was another Baldwin sheep pasture until the late 1880s when E.J.'s cousin, Charles Baldwin, with his wife Julia, took over supervision of the acreage and improved it into "one of the leading dairy ranches of Southern California."[104]

La Cienega, as well as parts of other Baldwin Ranch lands, was to flow in oil in the twentieth century, and though the prospect of E.J. as an oil tycoon is intriguing, such was not to be. The acreage remained only ranch land throughout Baldwin's lifetime.

Baldwin's Famous Old Prize Wines and Brandies

These wines and brandies are made on the world-renowned Santa Anita Rancho, and are sold direct from the wine vaults only, and are the oldest wines in California.

PRICE LIST

	Per case of 12 qts.
Old Port 20 years old	$ 10.00
" Muscat " " "	10.00
" Sherry " " "	10.00
" Angelica " " "	10.00
" Madeira " " "	10.00
" Tokay " " "	9.00
" Grape Brandy, 12 years old	20.00
Apricot Brandy Cordial, 15 years old	15.00
Orange Brandy Cordial, 15 years old	15.00

	Per case of 12 qts.	Pts.
Zinfandel, 7 years old	$ 7.50	$ 8.50
Burgundy (Superior), 9 years old	8.00	9.00
Sauterne (Santa Anita), Dry, 10 years old	8.00	9.00
Chateau Yquem, 10 years old	10.00	11.00
Olive Oil	12.00	
Olives (Ripe)		

Freight rate to ....$........per 100 lbs. or 2 cases.

All orders carefully packed and shipped to any part of the United States or Canada. Address all orders to

M. LAWRENCE & CO., Sole Agents,
Arcadia, Los Angeles Co., California.

Chapter Four

Land Poor and Short On Luck

Lucky Baldwin spent liberally to develop, improve, and advertise a host of California projects in his post-Comstock days, and it was his Southland ranch that both supported and suffered from the financial commitments therein engendered. "It is true, I am getting old and could afford to take things easy," Baldwin said, "but this is not my nature. . . . So long as I have money I will keep spending it on improvements."[1]

Still popularly acclaimed a millionaire, and perhaps so in his own subconscious, Lucky Baldwin nevertheless spent borrowed money for the last thirty years of his life. The mortgage effected in 1878, in the aftermath of his $1.5 million stock loss, was never fully paid off. Interest payments and land taxes were regularly met, but all remaining cash was continually put back into his investments.[2] By 1880, E.J.'s wealth was land-based. When real estate values were high, Lucky Baldwin was lucky indeed, but when depression struck, as it did with great regularity in the nineteenth century, Lucky was left land poor.

The Southern California boom of the eighties had provided a respite for Baldwin. The early Sierra Madre and Santa Anita Tract sales allowed E.J. to participate with capital as well as with enthusiasm in the intrigues of town building, but as the boom "gradually shrivelled up,"[3] so, too, did Baldwin's cash flow. E.J. had been land poor before the boom and he became so again with its demise, but the ensuing 1890s were a decade of national depression as well, and Baldwin's long-standing financial difficulties were accordingly exacerbated. In 1896, "Baldwin's Big Mortgage" was headlined in the *Los Angeles Times*. "The heavy mortgage effected by E.J. Baldwin with the Hibernia Savings and Loan Society," it was explained, "was not a renewal of a mortgage, nor was it an original one, but rather a concentration of mortgages heretofore held by individuals at various rates of interest."[4] With the exception of the acreage immediately surrounding the Santa Anita homesite, all of Baldwin's Los Angeles and San Francisco properties were tendered as security on the blanket mortgage," a $1,625,000 debt bearing 6.5 percent annual interest.[5]

Baldwin had fixed costs to meet—taxes and interest payments, specifically—that could not be postponed, but he also continued and even initiated projects in the 1890s that one could label incomprehensible if not familiar with E.J.'s policy of constant investment. "I did not need to take such a big mortgage as I have just taken on my property," he said, "but I like to have plenty of money to improve with."[6] The methodology was succinctly expressed by Jackson Graves. "Lucky Baldwin," he wrote, "always bored with a big auger. As a consequence, he was always in debt."[7]

The money invested in Santa Anita proved to be well spent. "Income from this tract of land," remarked Hiram Unruh in hindsight, "has been the main dependence in taking care of losses, interest, improvements, taxes, etc. during the past twenty-five years."[8] The years were not so kind to other Baldwin investments, however, and the cumulative financial pressures of the 1890s laid bare the fragility of Baldwin's economic position.

Lucky Baldwin did not settle permanently in Southern California until 1900 and, logically, many of his capital-draining projects were in the northern part of the state. As one follows E.J.'s peregrinations in thc newspapers of the period, a pattern emerges of a man in constant motion from his Santa Anita Ranch, to the Baldwin Hotel in San Francisco, to Tallac, his Lake Tahoe resort.

Not all of Baldwin's projects lost money, but both his San Francisco hotel and Tallac were expensive undertakings that ultimately brought E.J. more prestige than income. The earnings of the Baldwin Hotel, in fact, did not even pay expenses under Baldwin's initial management.[9] Jim Marvin, chief clerk at the Baldwin Hotel from 1877 to 1882, remembered that E.J. "was mighty hard up, though he owned thousands of acres of the finest land in the state." The hotel lost money, according to Marvin, because "we gave too much for our rates, and the overhead was too high, and there was too much competition from the Palace."[10]

The Baldwin Hotel was begun in 1873 with profits from E.J.'s Crown Point mining shares, and it was completed five years later with a $3 million price tag.[11] With a good location at Market and Powell Streets, commanding five stories in height and a quarter mile frontage,[12] the Baldwin Hotel was a luxurious ego-booster for E.J. Furnishings were acquired at the Centennial Exposition in Philadelphia to outfit four hundred rooms, plus the bar and restaurant, the office, and reading and billiard rooms; a $25,000 Tiffany clock set the tone for an interior decorating scheme that took a million Baldwin dollars to execute.[13] Nor was that all. Within the hotel structure was Baldwin's ornate Academy of

Baldwin Hotel Dining Room

Music, a theater, in the words of Bancroft, "acknowledged to be the most comfortable and tastefully embellished west of Chicago."[14]

Baldwin loved the theater and lured several great actors to his Academy of Music over the years, but artistic success was not necessarily measured in cash profits, and quite early on E.J. tried to ease the drain on his resources by renting his hotel to an outside party.[15] Henry Pearson signed a five-year, $3,000 a month lease in December of 1882, to free Baldwin for other projects.[16]

The hotel business was something Baldwin knew and enjoyed. In November, 1880, he purchased some eighteen hundred acres of virgin timberland, one mile fronting on Lake Tahoe, the balance stretching south to Fallen Leaf Lake, and with the acreage he acquired "Yank's," a forty-room hotel already on the property.[17] "Yank's" became Baldwin's Hotel Tallac (named for Mt. Tallac, which overlooked the property) and in1881, E.J. leased the hotel to Melville Lawrence and George Comstock.

Baldwin, Lawrence, and Comstock (George died in 1888 and was succeeded by his son, Harry) formed a potent combination who methodically developed the resort potential of the property, first with walks and bridle paths, later with the addition of a lake steamer, and finally, in 1898, with a "palatial three and one-half story edifice . . . known in grand simplicity as 'The Tallac.'"[18]

E.J. borrowed $30,000 to help finance "The Tallac," but business seemed brisk, and two years later a matching "Casino" was constructed nearby "for the amusement of guests."[19] Tallac

Tahoe Tallac Hotel

and its Casino were billed "the summer resort of the world" by Lawrence and Comstock (who similarly ran E.J.'s Hotel Oakwood in Arcadia as a winter resort), and visitors were treated to not only the pleasures of the outdoors, but also to the indoor recreations of pool, billiards, dancing, bowling, theatricals and, when the sheriff was safely in Placerville, surreptitious gambling.[20]

In the absence of documentation, it is difficult to assess the overall profitability of Baldwin's hotel ventures. The likelihood, though, is that E.J. typically plowed his profits into improvements in hopes of an ever brighter future while continuing to rely on Santa Anita as the pocketbook for his empire.

A development project that used Baldwin capital but produced no return was E.J.'s Bear Valley acreage in Southern California. Baldwin had acquired, through purchase and state grants,[21] some six thousand acres of land in the vicinity of the Gold Mountain Mine (including what is today known as Baldwin Lake) and in 1890 it was reported that plans had been advanced for "a large summer hotel" to be built near a mineral spring on the property.[22] The hotel plan was soon either replaced by or became part of a larger Baldwin scheme designed "to take the waters of Van Dusen Canyon and the second lake in Bear Valley (Baldwin Lake), and by means of a tunnel about two miles long put them on the other side of the mountain down upon what is now a desert."[23]

To initiate his irrigation project, Baldwin incorporated his own Bear Lake Land and Water Company in 1891[24] and announced his determination to contest the right of the powerful Bear Valley Improvement Company to mutually-claimed Van Dusen waters. Bear Valley Improvement was building the "big reservoir" at Big Bear, however, and their claim to the valley waters apparently took precedence over E.J.'s.[25] A deed conveying Baldwin lands to his Bear Lake Land and Water Company (July 2, 1891) carries the hand-written notation, "not recorded, company organization not completed."[26]

Neither Baldwin's Gold Mountain Mine nor his irrigation project, then, bore fruit in Bear Valley; son-in-law Budd Doble (Clara's third husband) was subsequently unsuccessful when he attempted to run the mine in 1893-95,[27] and before the turn of the century both the mine and a good deal of the Bear Valley acreage were disposed of by a surely disappointed Lucky Baldwin.[28]

More satisfaction, though it too carried a price tag, came with Baldwin's notorious love life. On May 19, 1884, Elias Jackson Baldwin ignored a forty year age difference to make sixteen- year-old Lillie Bennett his fourth—and final—wife.[29] "Lillie was a beautiful girl," recalled a childhood friend, "and Baldwin proceeded to fall in love with her, rushing her into marriage before she had time to think everything through."[30]

All was well at first. "For a year after she married Baldwin, this little girl was queen of the ranch," wrote the *Times*. "Then," however, "came a change; the ever-fickle Lucky found a new 'affinity.'"[31] Rather than divorce her philandering husband, Lillie, with the aid of her mother, retreated to Northern California and "made Baldwin come to time" with a "magnificent home in San Francisco" and "orders to the management of his Baldwin Hotel to look after her every need."[32] E.J. continued to support his wife throughout his lifetime, and Lillie in turn "was always ready to welcome" her husband when he chose to visit his San Francisco home. When Mrs. Baldwin visited Santa Anita, "which was at long intervals," according to the *Times*, "she absolutely ignored everybody that bore any semblance to an 'affinity.'"[33]

Her determination was to be admired, for Baldwin "affinities" were legendary, perhaps the best publicized aspect of E.J.'s personal as well as financial life. The costs incurred in the courting and uncourting of a parade of young females, in fact, became a regular expenditure for Lucky, a unique, but nevertheless authentic drain on his financial resources. Three of Baldwin's "affinities" took him to court, one just prior to his marriage to Lillie Bennett, one just after, and the third when E.J. was sixty-five years old.

Friend and foe alike agreed that Lucky Baldwin readily and openly defied social conventions. "He revealed no moral or spiritual scruples or evidence of a painful conscience," wrote Glasscock.[34] In one law suit, E.J. even advanced the theory that "his reputation was such that every woman who came near him must have been warned against him in advance."[35]

The stories of "Lucky's Litigious Loves"[36]—Verona Baldwin, 1883; Louise Perkins, 1886; and Lillian Ashley, 1893—can be found in detail in Glasscock's biography, in short articles that borrow heavily from Glasscock, and in amazingly graphic news reports of the period. Los Angeles' Horace Bell used his sheet, *The Porcupine*, in relentless assault on "Beast Baldwin," accusing him not only of numerous seductions, but of bribery in attempts to defend himself.[37] "We want to see the Beast punished by the only way such vile creatures can be," wrote Bell.[38] An "angry Baldwin," in turn, filed sixteen counts of libel against Horace Bell, "alleging damages of $5000 in each,"[39] and the *Los Angeles Herald* reflected the confrontation by publishing seventeen straight days of word-by-word testimony in the Louise Perkins breach-of-promise trial.[40]

Lucky Baldwin settled the Louise Perkins and Verona Baldwin suits out of court and defeated Lillian Ashley, but not before his reputation as a Lothario was firmly entrenched, and not without the expenditure of thousands of precious dollars. E.J. employed fourteen lawyers in the Lillian Ashley suit,[41] and the bill from Lloyd and Wood alone was $11,000.[42] In a later interview, Lucky would recall the law suit "that cost me $200,000."[43]

Raising the cash to meet his financial obligations was a constant struggle for E.J. His plan of survival was three-fold. He borrowed, and accepted interest payments as a by-product; he invested, and ran the chance of loss; and he conserved. He did not, as popularly suspected, gamble for profit. A publicity stunt at the Baldwin Hotel, in which piles of $20 gold pieces were arranged on a poker table in front of him, perhaps started the rumors that Lucky Baldwin was an inveterate gambler, and his reputation no doubt made such rumors plausible, but, according to his biographer, E.J. played cards for small stakes, "and he played them close."[44]

Louise Perkins

EJB and Friends at Poker in Private Quarters, Baldwin Hotel

"Though he loved a poker game as much as he loved women," wrote Glasscock, "he was no more generous in one than in the other."[45]

On a less personal note, Baldwin was generous with both love and money with his race horses, though the pressures of the 1890s took a toll on the Baldwin stables as surely as in other of E.J.'s enterprises. In 1890, Lucky announced through the newspapers that he would leave the turf at the conclusion of the current racing season. "I will not retire without regretting the step," he said, "for I take great pleasure in the raising and racing of thoroughbreds but my business enterprises in California will for the next few years engross my entire attention."[46]

Baldwin informed buyers that he would sell all the yearlings and two-year-olds at his Santa Anita Ranch, then continue breeding with an eye to an annual sale of stock, but almost before the ink had dried on the presses, E.J. changed his mind.[47] Baldwin racers continued to compete until 1898 when "the owner of the far-famed Santa Anita stable announced that he would never again ship a stable across the Rocky Mountains."[48] The decision lasted a year longer than E.J.'s earlier one, but the attack of economy passed quickly, and the *Times* reported that Lucky Baldwin would be at Saratoga in 1900 "with a formidable stable."[49] Until the day of his death, Baldwin thoroughbreds were active in the turf world.

Lucky Baldwin's economizing found more direct outlet among his employees. There were over one hundred men on the ranch payroll, which in 1883 had been reported as averaging $4000 a month,[50] but E.J. was said to pay wages "more regularly with promises than with cash."[51] According to Hiram Unruh, "at times as much as $40,000 was owing to employees."[52]

In 1889 Charles Anderson was hired as bookkeeper and, with Baldwin's blessings, devised a system of issuing trade orders on the ranch store to pass as currency. Regular paydays ceased, and employees were paid in cash only infrequently, receiving instead charge accounts at the store.[53] Even higher-salaried employees, such as Charlie Anderson, were likely to find their wages in arrears, and if the amount went too high, an acreage lot in one of E.J.'s subdivisions was offered in lieu of salary.[54]

Jobs were in short supply nationally in the 1890s, and, in the words of a Baldwin associate, E.J. "kept dozens of families and dozens of single men on his ranches with shelter, food, and clothing during all those hard times." Workers were "virtually in a state of peonage," he agreed, but quickly iterated that "a peon on the Baldwin ranch was much better off than a free man on the streets in those days."[55]

"One day Mr. Baldwin and I were riding down a lane along which a lot of peacocks were perched on fences uttering their shrill cries," reminisced Hiram Unruh. "Mr. Baldwin said, 'Just listen to that. Even the peacocks are yelling "Pay off!"'"[56] As his financial obligations mounted and his cash income dwindled, E.J. gained a well-earned reputation of rarely paying a debt without a court judgment and only infrequently paying a court settlement without an appeal.[57] Even a partial list of Baldwin law suits of the period is formidable and of surprising variety. Boundary and water disputes were complex, expensive, and frequent; litigation concerning spurned women carried on for years; and everyone from lawyers and commissioned artists to equipment salesmen and employees sued E.J. for delinquent payment of bills.[58] A friend once asked Baldwin why he allowed so many suits to be brought and why he appealed almost all judgments. Lucky replied,

> Everyone in the country, including the women, thinks I have millions, and they're all ready to rook me as an easy mark. This is the way I prove they're wrong, and maybe discourage their efforts. I want a reputation of being hard to collect from. If anybody wants anything out of me for any purpose, they'll have to sue.[59]

Lucky Baldwin was tough, and he would have to be to withstand the final blow of the 1890s. In the early morning hours of November 23, 1898, the Baldwin Hotel, E.J.'s three million dollar San Francisco showplace, caught fire and was rapidly burned beyond salvage.[60] Despite extraordinary fire precautions that had included two artesian wells in the basement plus two additional water tanks on the roof of the hotel,[61] the "Baldwin" had a vulnerable wooden interior and burned into "a mass of smoking ruins" within two hours of the first alarm. The fire,

EJB with Employees

which started in the hotel kitchen and was quickly sucked up to the roof through an elevator shaft, claimed four lives among the three hundred registered guests, and observers expressed wonder that so many escaped unharmed.[62]

Grieving, but physically untouched by the conflagration, Lucky Baldwin was left instead to calculate his financial loss. "The insurance," E.J. told reporters, "was light in comparison to the loss," and indeed, total coverage was reported as a meager $185,000.[63] Even as the figures were cited, creditors attached liens on the insurance money, and Baldwin was pressed for information on his plans for the site.[64]

"He didn't molt a feather," commented Byron Waters, one of E.J.'s attorneys. "His courage never faltered. He never to the day of his death lost his enthusiasm. And even on that day he was planning what he would do on the next."[65] In the days immediately following the fire, Baldwin weighed the difficulty of selling his heavily mortgaged hotel site against his desperate need for cash. He contemplated erecting a row of stores on the unimpaired hotel foundation;[66] he even spoke of forming a corporation to build a new hotel.[67] In 1898 he did what his financial plight required, however, and sold the Market and Powell Street property (excluding the adjoining, but less valuable "Baldwin Annex") to James L. Flood for $1.1 million.[68] The purchase price was $200,000 above the mortgage held on

Burning of Baldwin Hotel

the site and gave Baldwin some capital to begin the process of recouping his many losses.

"If Baldwin was no longer lucky," commented the *Los Angeles Times*, "he was at least plucky."[69] In the summer of 1900, at the age of seventy-two, Lucky Baldwin set sail for Nome, Alaska. He was accompanied by David Unruh, Hiram's son, who stressed that for Baldwin "this was a business venture, inspired in part by a desire to recoup his fortunes in the Alaska gold rush, and in part by a desire to renew his youth. . . . He hoped to locate or buy some promising land claims," said David, "but he had no illusions about mining."[70]

E.J. was a businessman, and it was not a pick and shovel he transported to Alaska, but rather two ready-cut frame houses, gambling equipment, and "a considerable quantity of liquors for his proposed bar."[71] Businesses proved as difficult to establish as Alaskan mine claims, however, and the lot David finally secured was too far from the center of town to be profitable. "The gambling outfit was never even uncrated," said the young Unruh. "We went out on the last boat that fall," leaving a caretaker.[72]

E.J. was ill when he and David started back in the spring. A young woman who accompanied the men, rumored to be an adventuress, was instead a nurse, "and a good thing she was, too," said Unruh. "By the time we reached Seattle the old man was so bad off he had to be taken off the boat and she went back to California with him by train." [73] David Unruh went on alone to Alaska to close out E.J.'s business interests,[74] and a weak and crotchety Lucky Baldwin was taken to his Southern California ranch to recuperate. Fearing the worst, concerned family and friends followed, but as a *Times* reporter cautioned,

> It must not be inferred that Mr. Baldwin has any notion of "passing in his checks" or relinquishing his grip on his vast possessions very soon. He is of the sturdy type of pioneers who never say "die," and will probably be found in harness when the final summons comes.[75]

Chapter Five

City-Making—Methods and Motives

Lucky Baldwin did regain his health amid the quiet of his Santa Anita home, and he decided that it would be in the Southland that he would recoup his lost fortune. The national economy was improving with the turn into the new century, and as land began to regain its potential as a salable commodity, E.J. prepared to reap the benefits. Santa Anita became Lucky's permanent home in 1901; "he says there is Nome enough for him in Southern California," reported the *Times*.[1]

His ranch, however, had suffered in his absence. A series of dry years had cut crop production on his outlying acreage, and even at Santa Anita unpaid employees threatened strikes and work slowdowns. "But when Baldwin returned from Alaska and took hold of his tangled affairs here," commented a reporter, "Presto! In spite of dire predictions, there were no foreclosures of mortgages. Instead of that, debts were paid off at a rapid rate . . . and he has revolutionized things at the ranch."[2] Santa Anita was returned to its high standards of beauty and productivity as part of E.J.'s improvement campaign, and neglected outlying ranches were given increased attention. Baldwin, it was observed, "set himself assiduously to the task of making his vast farm . . . paying property."[3]

Lucky Baldwin relished the role of land owner and generally sold property only when pressed. He was a "good holder," in Hiram Unruh's words,[4] but the need for cash was a constant factor in Baldwin's financial empire, and as land values rose with an improving economy, E.J. began a new series of subdivisions on his properties. The Spring Street frontage in Los Angeles had long provided a steady rental income for Baldwin, and in 1904, another non-ranch parcel, his eighty-acre "Lucky Baldwin Tract," was subdivided and put on the market.[5]

Rancho La Cienega, according to Unruh, was "ripe for subdivision," too, with its prime location adjacent to the western limits of the city of Los Angeles.[6] The Redondo Electric Railway and the Santa Monica branch of the Southern Pacific traversed the property to increase its value and fuel "a demand for city lots and residence sites," said Unruh, but although he estimated in 1908 that the value of the thirty-five hundred acre ranch had already risen to $7 million,[7] E.J. was never convinced to part with his prized acquisition. Until his death, La Cienega was run as a model dairy producing a respectable $25,000 annual income.[8]

In the San Gabriel Valley, Baldwin and Unruh similarly held the majority of ranch lands in expectation of increasing values. Subdivisions were made as cash was needed, but not before, for, as Unruh explained in a mortgage request, "the value of the property is going up quite rapidly and E.J. doesn't want to sell now. With a loan, he can wait three or four years, improve the property, then sell at a much better price."[9]

Though a suspicion lurks that Lucky was simply loathe to part with lands he had spent twenty-five years shielding from foreclosure, the logic of "improve and sell" is not only irrefutable, but consistent with long-standing Baldwin business practices. Among Valley lands, Rancho San Francisquito had become increasingly suburbanized over the years as E.J. sold lots piecemeal to meet various obligations. In 1908, only one-sixth of its original six thousand acres remained in Baldwin's hands, some of it rented for strawberry growing and the remainder used to grow feed for Baldwin livestock.[10] Rancho La Merced in its entirety and most of the Ranchos Felipe Lugo and Potrero Grande, on the other hand, continued in use as pasture land[11] though the 1886 subdivision of Rancho Felipe Lugo (fifteen hundred acres) had been adjoined in 1898 by "E.J. Baldwin's First[12] and Second Subdivisons in Rancho Potrero Grande."[13]

The huge Rancho La Puente was variously utilized by Baldwin at the turn of the century. Some six thousand acres of it were "farmed and rented" for hay and grain cultivation and for sheep raising, while another eight thousand acres was "rented on shares for beans, potatoes, hay, and grain."[14] Parts of the remaining acreage, though, were carved into five different subdivisions by Lucky Baldwin between 1904 and 1907.[15] The first plat recorded, "El Monte Walnut Place," covered a large part of today's Baldwin Park (south of Ramona Boulevard),[16] and was, in fact, indirectly responsible for the christening of E.J.'s namesake town.

Early in 1906, rumors were being bruited about that Baldwin was planning a townsite on his subdivision south of the new Pacific Electric right-of-way[17] and just southwest of the old settlement of Vineland.[18] E.J.'s town, it was said, would be called "Baldwinsville," and would no doubt quickly eclipse the less prestigious Vineland. Dewey Shultis, a Vineland shop keeper, and a group of concerned local diplomats approached Baldwin with a proposal suggesting that in lieu of competing business centers, the "Baldwin" name simply be transferred to their town.[19]

E.J. agreed to visit the Shultis store to discuss the matter, and in a now legendary story, no more than put his foot on the makeshift box serving as the store's doorstep than he toppled over backward. Luckily, the story goes, Baldwin fell into the arms of Mrs. Shultis behind him, and perhaps in gratitude, though more likely because "Baldwinsville" had not been much more than rumor in the first place, E.J. agreed to the use of his name for the Vineland/El Monte Walnut Place settlement. Dewey Shultis, delighted, hastily painted "Baldwin Park" across his store front and, so christened, the town has remained.[20]

A bigger promotion and an uncontestedly more colorful tale surrounds Baldwin's Arcadia townsite. E.J.'s plan to nurture his ranch lands into "paying property" utilized various avenues of approach, but none as flamboyant or as personally attended as his brazen scheme to launch Arcadia into cityhood. The turn of the century found Arcadia as neglected as Baldwin's other Southern California properties, "a puny village of 360 souls,"[21] spread thinly over twelve square miles. "It is stated by responsible persons," said the *Times*, "that the land is adorned by not more than sixty-five buildings, some of which are barns, stables, and shacks, built of such material as cast-off water pipe, refuse tin, flattened-out gasoline cans and dilapidated shingles."[22] Hardly a propitious starting point, but Lucky Baldwin was a determined man.

It was the Pacific Electric Railway that forecast the rebirth of Arcadia, just as it keyed a suburban revival throughout the Southland at the turn of the century. "The electric roads," commented a contemporary observer, "are bringing the whole country-side within touch of the city, and as a direct result there are such rapid increases in values as makes one's head swim."[23] In 1902 it was announced that Henry Huntington's Pacific Electric would soon start construction of "a line from Oneonta, on the Pasadena short line, through Arcadia to California Street in Monrovia."[24]

With visions of inflated land values, E.J. signed an "agreement" with the P.E. in November of 1902[25] and on March 1, 1903, happily watched the red cars pass through his Arcadia townsite.[26] The stage was set and two weeks later, in a surely calculated sequence, Lucky Baldwin filed a petition for the incorporation of Arcadia into a municipality of the sixth class.[27] "I wish to incorporate a city to enhance my property and that of the surrounding district," Baldwin stated,[28] but neighboring towns argued that Baldwin's explanation was only the tip of a decidedly objectionable iceberg.

Opposition to the birth of Arcadia was immediate and vociferous. Under the legal provisions of the Municipal Corporations Act of 1883, incorporation required a verified city population of at least five hundred, plus submission of an official petition to the County Board of Supervisors describing proposed city boundaries and specifying the number of inhabitants.[29] Public objections were to be heard and filed by the Board of Supervisors prior to the incorporation election and, as the *Los Angeles Times* predicted, "Lucky Baldwin's city" became "a live issue."[30]

Protestors charged that Baldwin had personally, artificially, and illegally inflated Arcadia's population to reach the number required for incorporation. "By the help of Huntington's illiterate peon gang, which was shipped out there, he was able at the time of the census taker's visit to show 642," one opponent claimed,[31] while the *Times* more generally commented that if Arcadia had five hundred inhabi-

Pacific Electric Cars

tants, "a very considerable proportion of the population is composed of members of a railroad grading camp at present encamped on the land, and another considerable proportion is made up of Baldwin ranch hired hands."[32]

A third variation of the inflated population charge came from Edward Hahn, attorney for the opposition, who presented hearsay evidence to the Board that some one hundred laborers had been hired at a Los Angeles employment agency to go out to Santa Anita for a single day. They received a dollar in cash, car fare, and a free dinner, said Hahn, in return for permitting their heads to be counted by the census taker.[33] Locally, in yet another version, Charles Jones, mayor of Sierra Madre in 1907, remembered that Baldwin simply gathered all the children on his ranch and had them quietly transported from settlement to settlement to be counted, then recounted in the census.[34]

In truth, though, incorporation opponents were less concerned with Lucky Baldwin's methods than with his motives. Whether E.J. colluded with Huntington, counted temporary residents, hired temporary residents, or simply manipulated ranch employees, the fear was that with incorporation Lucky would be free to initiate what was labelled "a black tide of damnation."[35]

Opposition to the birth of Arcadia was centered in the Anti-Saloon League, who charged that E.J. sought incorporation "simply and solely for the purpose of establishing in the heart of Los Angeles County an American Monte Carlo, with whose ribaldry, racing, gambling, and gaming the county government would be powerless to interfere."[36] Dr. Chapman, president of the Anti-Saloon League, wrote to "each of the 160 pastors in Los Angeles County" urging them to circulate petitions among their congregations protesting Baldwin's machinations as a "gambler's scheme" that would "produce widespread harm to the morals, business interests, and good name of Southern California."[37]

Dr. Chapman confronted Lucky Baldwin face-to-face in scheduled May hearings before the Board of Supervisors, but the wily E.J. "denied absolutely that he intended turning Arcadia into a gambling hell," refuted charges of population inflation,[38] and asserted that with incorporation he intended only to improve his property, provide for a local school for Arcadia residents, and, incidentally, relieve himself of a $2500 yearly tax on county roads that traversed his Arcadia acreage.[39]

The opposition clearly disbelieved Baldwin, but when questioned by Lucky's attorneys, Dr. Chapman admitted that he could provide no evidence of the "Monte Carlo" possibilities of Arcadia "except the gambling habits of the petitioners."[40] The population charges were similarly difficult to verify for the protestors, and, though remonstrances were legally recorded, the Board of Supervisors unhesitatingly approved Lucky Baldwin's Arcadia petition and authorized an incorporation election to be held in the town on July 27, 1903.[41]

Thirty-nine voters turned out to select cityhood for Arcadia and to cast ballots for a five-man Board of Trustees, a clerk, a treasurer, and a marshal for the city. There was "nary a 'nay'" among the thirty-five votes cast in favor of incorporation, reported the *Los Angeles Times*, nor was there particular agitation over the election of city officials.[42] Both the *Times*[43] and the *Pasadena Evening Star*[44] ran a complete list of the men who voted in Arcadia's first election, and fully 75 percent of the names are readily identifiable as Baldwin relatives, business associates, and/or employees.[45]

The slate of candidates voted into office included all but one member of the election board, and all, in the words of the *Times*, were "faithful subjects of the King of Arcadia."[46] Elias Jackson Baldwin, with thirty-eight votes, became mayor of the new municipality, and in descending order of votes received, the other four board members were Melville Lawrence, leasee of the Hotel Oakwood; H.A. Unruh, Baldwin's faithful manager; D.P. Unruh, Hiram's son and E.J.'s shipmate to Alaska; and Hull McClaughry, Anita Baldwin's husband and Lucky's son-in-law.[47] To complete the monopoly, Charles and Elmer Anderson, Baldwin Ranch foremen, were elected respectively city treasurer and city marshal, and employee J.P. Cavallier, also a former Baldwin business associate, became city clerk.[48]

Lucky Baldwin's explanation of his brash entry into city-making was typically enigmatic. "I found," he said, "that I couldn't live like a free American out here unless I went into politics."[49]

Neighboring Sierra Madrans viewed Baldwin's politics as spiteful; Baldwin lost three water suits to Sierra Madre, said Mayor Charles Jones, and in retaliation he incorporated Arcadia acreage to prevent Sierra Madre from ever expanding beyond its southern and eastern boundaries.[50] A more likely motivation, however, can be found in the intriguing comment of the *Los Angeles Express* that "in voting for Mayor Baldwin the denizens of Arcadia may be said also to voice a protest against the proposed annexation of the sylvan place to Greater Los Angeles."[51]

In 1903, as part of an ongoing effort to acquire a harbor in San Pedro Bay, the City and County of Los Angeles appointed a joint commission to investigate the possibility of a merger between the two government units.[52] Consolidation was never effected, but E.J.'s disdain for Los Angeles, where, he claimed, "you have to pay a tax to have a trunk moved across the street,"[53] was such that the mere threat of rule by city ordinance could have propelled him into his incorporation scheme. More to the point, gambling was illegal in the city of Los Angeles,[54] and though he had flatly denied before the Board of Supervisors "any intention of establishing a town for the fostering of horse racing, gambling, and kindred vices,"[55] Lucky Baldwin belied his words in an interview with the *Los Angeles Times* on the day of incorporation of his Arcadia townsite.

Arcadia was E.J.'s plaything, in many ways. It was incorporated in a scheme of Baldwinesque audacity, and it would reflect its maker's interests in the years immediately ahead. "We are more certain than ever that the place is to be made a gambling hell and booze pleasure park," opponents had charged,[56] and Baldwin, with "a glint in his eye and a rosy flush mounted to his temples," responded, "Damn it, we'll show 'em a race track and plenty of it, the long-haired kickers."[57]

In the autumn of 1902 the first meeting in Southern California entirely for running horses was held in Agricultural Park, and it proved so popular, promoters moved quickly to establish a track especially for thoroughbred racers.[58] If Lucky Baldwin was interested, and indications, though conjectural, are that he was, he did not move fast enough nor, probably, with sufficient cash in 1903 to secure the prospects of a race track for Arcadia. "He completed the surveys for his own plant and tucked the blueprints away for future use," said Glasscock,[59] but it was backers of Ascot Park who broke ground in early summer and on Christmas Day, 1903, opened a race track on unincorporated property at the corner of San Pedro and Slauson.[60]

Baldwin could not profitably compete with Ascot, but even without a race track, Arcadia was a city of unusual promise. When questioned "as to the Monte Carlo possibilities" of his city, E.J. replied, "They are fine. It is a temptation and I am not saying but that if the right man came along I might lease or sell the privileges, but I am an old man and have plenty to keep me busy."[61]

"We are going to have a town here and a schoolhouse and a paper that can talk back," said Baldwin,[62] and he carried through on his promises. On August 3, 1903, a new Arcadia City School District was authorized by the County Board of Supervisors to serve area students who had previously been attending schools in Monrovia, Sierra Madre, and Chapman,[63] and E.J., who appointed himself chairman of the school committee, immediately donated a ranch packing shed at the northeast corner of Santa Anita and Falling Leaf Avenues for use as Aracadia's first local school.[64]

The first school was joined by the first city newspaper on May 7, 1904, as the four-page weekly *Arcadia Bulletin* began its job of publicizing city events.[65] Church, library, banking, and club news was omitted from the *Bulletin's* pages, however, for there were no such city amenities in Lucky Baldwin's town.[66] Tourism and entertainment rather than banks and businesses were the cornerstones of early Arcadia, and a "Local Tid Bit" of December 24, 1904, encouragingly reported that "several tally-hos of people are beginning to come out from Los Angeles every week now to stop at the Oakwood and drive over Baldwin's ranch."[67]

One of the first ordinances passed by the Board of Trustees on September 9, 1903, was a provision "for the licensing for the sale of liquors" in the city of Arcadia,[68] and Clara Baldwin Stocker, E.J.'s elder daughter, filed for the first license on April 27, 1904.[69] Three additional applications were filed in 1905, and another the following year,[70] prompting a Pacific Electric travel magazine to report that "Arcadia, known as Baldwin's town, is a popular village and 'wide open' in its government."[71] Primarily, Arcadia

Hotel Oakwood with Tally Hos

was known for its entertainment," corroborated Julian Fisher,[72] and, indeed, visitors of all types were encouraged to enjoy Arcadian hospitality.

In the summer of 1905 the Pacific Electric initiated its "Orange Grove Route" for San Gabriel Valley excursionists,[73] and Lucky Baldwin, though forced by the influx to post guards around his valuable ranch land, nevertheless was presented with a welcomed escalation of property values. The Pacific Electric, wrote a contemporary observer, "gashed a path" through the orchards and groves of the Baldwin Ranch to bring "real estate speculators nibbling at its outskirts."[74] Ranch land could be sold in1906 for "more than fifteen hundred dollars an acre," the writer claimed, as fast as Baldwin was "willing to let it be chopped up by the city broker."[75]

The four thousand acres of the "Home Place" were never offered for sale by Baldwin, but the beauties of Santa Anita served admirably as a public attraction, a splendid advertisement for those lands E.J. did put on the market. On the "Orange Grove Route," the Pacific Electric's "Parlor Car Poppy" left downtown Sixth and Main daily at 9:30 a.m. to "take in the Ostrich Farm, San Gabriel Mission, Lucky Baldwin's Ranch, and Monrovia," stopping for an hour or more at each point of interest to allow excursionists time to enjoy the sights and activities of the areas.

For those so inclined, tally-hos met the cars in Arcadia to offer visitors, "for the nominal price of fifty cents," conducted tours of the Santa Anita Ranch.[76] "On this drive," it was promised, "all points of interest are visited—the house, winery, racetrack, and stables," and at its conclusion, all participants "are invited to sample the rare vintages at the Hotel Oakwood, free."[77]

"This is a tremendous big ranch," wrote one bemused tourist, and it is "owned by the old man Baldwin who has a notoriously bad reputation . . . but he is very rich as well as very bad."[78]

Wealth and reputation, though, did not shield "old man Baldwin" from the trauma of eye disease in his advancing years. E.J., it was reported, was sent to San Francisco in the fall of 1905 for a two-month program of treatment, and when he returned, "with the sight in one eye completely restored and with the inflammation in the other greatly reduced," it was to find his homesite—grounds, guest cottage, and adobe home—each "bravely illuminated with innumerable electric lights."[79] The view must have been encouraging. Nearing his seventy-eighth birthday, but with at least partial vision restored to him, Lucky Baldwin, it was written, once again exhibited "a zest that belied his years" as he embarked on one last round of Southland activities.[80]

Post-Rose Parade chariot racing in Pasadena's Tournament Park provided E.J. with an amusing pastime from 1906 to 1908, but watching driver C.C. West pilot Baldwin horses, however daringly, was no comparison to the thrills of thoroughbred racing for veteran turfman Lucky Baldwin.[81]

"Several years ago Lucky Baldwin first conceived the idea of a race course on the Santa Anita Ranch," commented the *Los Angeles Examiner*, "and after many ineffectual attempts to promote the scheme, the passing of Ascot threw the chance in Baldwin's way."[82] Ascot Park had been in operation only three years when it fell victim to annexation by Los Angeles—where racing was prohibited—as part of a "shoestring strip" that tied the city to harbor acquisition.[83] George Rose and John Brink, backers of Ascot, set out at once to interest wealthy men in building a new track, however, and they had not far to look. Barney Schreiber, James Holland, Franklin Randle, and Dr. J.S. Gardner were turfmen of means, and in Arcadia, Elias Baldwin offered the consortium a building site unencumbered by gambling restrictions.[84]

"Lucky Baldwin's dream of a little kingdom with nothing in it but hot sports and himself as king, is about to be realized," reported the *Los Angeles Times*.[85] On September 30, 1907, the Los Angeles Racing Association was incorporated to serve as the operating club for Santa Anita Park, with E.J. and Hiram Unruh joining Rose, Brink, Schreiber, Holland, Gardner, and Randle on the Board of Directors.[86]

The corporation purchased 151 acres of land from Baldwin (located at the site of today's Santa Anita Golf Course) at $1000 an acre[87] and immediately contracted with A.M. Allen, "the hurry order race track builder,"[88] for construction of what would be billed "the newest and most modernly equipped of all race courses west of the Missouri River."[89]

Lucky Baldwin's $151,000 land sale price was paid to him in one thousand shares of capital stock in the Los Angeles Racing Association, making him the largest single stockholder, plus a note for $51,000 in cash. Rose and James Holland advanced the operating funds necessary to keep the corporation afloat, but it was Lucky, as major stockholder, who was elected honorary president of the Racing Association, and it was he who presided over opening day ceremonies.[90]

Santa Anita Park, not surprisingly, was patterned after Ascot. The grandstand, the *Examiner* reported, "is an exact duplicate of the one at Ascot and even the man who sells roast beef sandwiches is in the same old spot."[91] There were improvements, however, and reports prophesied that Santa Anita "will surpass the old track in many important particulars," including electrically lighted stables, in-house accommodations for stable hands, and a "fully equipped hospital for the use of the jockeys and other employees of the track."[92]

The Pacific Electric built a spur track from its Monrovia line to Santa Anita Park, and Southern Pacific, which normally made only occasional stops at Arcadia, arranged to run two excursion trains a day "from the Arcade Depot direct to the rear of the grandstand at Santa Anita."[93] Both rails, forecasting peak attendance, offered premium rates on their round-trip specials, and neither lacked business.[94]

Santa Anita Park opened December 7, 1907,

Grandstand at Santa Anita Racetrack

SPORTING SECTION.

Los Angeles Sunday Times

Part VIII

FORTUNES ARE LOST OR WON AT ARCADIA.

Mark Anthony II, Added Starter and Long-Shot, Wins Feature Race on Opening Day at Baldwin's Track—Thousands Wagered.

ALL-STAR COLLEGIANS DEFEAT ARMORY TEAM.

Los Angeles Times *Headline, Racetrack Opening*

with seven thousand patrons, several visiting dignitaries, and scores of reporters.[95] Noted the *Times*,

> Old Lucky Baldwin soared to the Seventh Heaven of seraphic bliss. At last he had a race track in his back yard. The Emperor of Arcadia stood at the top of the grand stairway in the grandstand, arrayed in a long-tailed black coat and a black slouch hat pulled down picturesquely over one eye. He looked out over the field like Wellington at Waterloo with the exception that a beautiful array of blonde damsels surged and frou-froued and billowed about him. [96]

Having suffered many reverses and achieved many successes in his nearly eighty years, wrote the *San Francisco Call*,

> Lucky Baldwin has now realized the greatest ambition of his life. "I desire no other monument," said Baldwin when the first day's races at his new track had been run. "This is the greatest thing I have ever done, and I am satisfied."[97]

Arcadia businesses were satisfied, as well, with Mayor Baldwin's race track. Earlier prognostications of a "gambling hell and booze pleasure park" were carried out with flamboyance in what the *Times* described as the "tough little debauched municipality" of Arcadia.[98] "Lucky Baldwin has made every arrangement to entertain his racetrack guests right royally," commented the *Los Angeles Express*. "There are several bars in operation in the only sporting town of Southern California," and "slot machines and rooms where friendly poker games are indulged in" are rife.[99] By 1909 the *Times* reported that "eleven saloons are now running full blast and auto parties of sports and fast women make business rushing."[100]

Hotel Oakwood remained the best known and most heavily patronized place of entertainment in Arcadia. When Santa Anita Park was opened, "the old hostelry was refurnished at a cost of more than $25,000 and electrical gardens were installed in connection with it at an additional expense of $10,000.[101] Slot machines and roulette wheels were found in the lobby, the bar dispensed Baldwin wines and brandies,[102] and in the rear of the hotel, for those interested, there was a boxing ring where some of "the fistic greats of yesteryear," including Jim Jeffries, Joe

Good morning
Much ado
Arcadia Station
Home of '49
The Boat Landing
Solid Comfort

Gans, and Battling Nelson trained and on weekends staged demonstration prize fights.[103]

A block north of the Oakwood, at First and St. Joseph, Clara and Harold Stocker's "Clara Villa" drew a mixed crowd of patrons that ranged from those who enjoyed the rooftop beer garden, to bowlers who used the inn's two alleys,[104] to those who, in Julian Fisher's words, patronized "the cribs, where the girls did their business." [105]

The "Norfolk," on Falling Leaf (Huntington Drive) near today's Colorado Place, catered to visitors who could not afford the Oakwood's prices[106] and so, too, did "Johnson's Inn," a "one-room beer place" at Santa Anita and White Oak (Foothill Boulevard), "Ben Newman's" at Falling Leaf and Second, "Mullin's" on First Avenue just north of the Santa Fe tracks, and the "Green Front Saloon" on the northwest corner of Santa Anita and Duarte County Road.[107] "White City," George Purvis's establishment on Santa Anita just south of the tracks, was widely advertised not only for its bar, but for its special weekend entertainments as well. "Baseball, balloon ascensions, high wire acts, a merry-go-round, and other attractions" drew Sunday crowds, while "every evening," it was promised, one could enjoy the Baldwin Music Hall—The Best in Musical Comedy and Vaudeville."[108]

Arcadia was a "sporting town," *par excellence.*

Clara Baldwin, Harold Stocker, and Friends Playing Croquet, 1894

Chapter Six

Death and Estate Settlement

"Lucky is a remarkable old man. One would think him to be twenty," commented the *Los Angeles Times*, "to hear him talk about his track and his town; his imperial dominion."[1] E.J. clearly enjoyed his last years, but he was not immune to the debilities of age, and in the early months of 1909 he contracted influenza which developed rapidly into pneumonia. He fought death with the vigor he had used to meet all life's obstacles, but his heart had weakened over the years, and "all but Lucky Baldwin," said the *Examiner*, "believed his end was near."[2]

On March 1, 1909, surrounded by family and friends, eighty-one year old Elias Jackson "Lucky" Baldwin died in the bedroom of his adobe home.[3]

Racing was suspended at Santa Anita on the following day to honor the Father of Arcadia, while at the homesite mourners quietly saluted the memory of the man who had so successfully turned his Southland property into a showcase ranch. On March 4, rancher and city-maker Lucky Baldwin was laid to rest in Cypress Lawn cemetery, just south of San Francisco.[4]

An era was passing, both on the Baldwin Ranch and in Arcadia as well. On April 7, 1909, it was officially announced by the management of Santa Anita Park that the track would close its meeting on April 17, "just two days in advance of the beginning of the operation of the Walker-Otis anti-racing law."[5] Horse racing was banned throughout the state in 1909, and as promoters sought hospitality elsewhere, it was prophesied that the "picturesqueness of Arcadia" would fade with the consequent departure of the "tough part of the population."[6]

"The closing of Santa Anita Park means that it [Arcadia] has sown its wild oats and must reform," editorialized the *Los Angeles Times*,[7] and the pronouncement was lent drama when just a week later, the colorful White City saloon was gutted by fire.[8] Hotel Oakwood burned on November 26, 1911, the day a new Sunday-closing ordinance for saloons went into effect in Arcadia. "There are those," said the *Times*, "who believe that the fire may have been part of a programme of revenge against the members of the Baldwin estate who . . . have been attempting to 'reform' Arcadia."[9]

The pattern of reform and revenge was repeated in 1912 when a vote to permanently outlaw liquor licensing in Arcadia was paired with the burning of the old racetrack grandstand, but the course had been set, and by 1914 Baldwin's town was well on its way to a new city image.[10]

Hiram Unruh was the mayor of Arcadia (1909-14) who presided over its transition to suburban tranquility [11] and it was he, too, who as executor oversaw the settlement of the Baldwin estate and its gradual transformation from farm land to profitable subdivision.

The Last Will and Testament of Elias J. Baldwin was notarized November 4, 1908, and when its terms were made public it was both applauded for its brevity and pronounced by attorneys "a masterpiece of legal handiwork . . . as near unbreakable as any document of its kind ever written."[12] Specific clauses in the will that elicited such praise stipulated that heirs who contested the document's provisions would be "cut off without a cent," while anyone not named in the will who managed to prove his right as an heir would be awarded a perfunctory sum of ten dollars.[13]

The clauses would be put to several well-publicized tests over the course of the four years needed to reach final settlement of the estate but, in 1909, initial public interest was focused on the naming of the beneficiaries of the Baldwin will and on the estimated value of the harvest they would reap. There were some surprises.

Comparatively minor and not unexpected rewards of La Puente acreage were bequeathed by E.J. to a niece and a nephew, Elizabeth Rush and Charles Fawcett, twenty acres each; to

Hotel Oakwood After Fire

Hiram Unruh, one hundred acres; and to granddaughter Rosebudd Doble Mullender, one hundred acres; but it was the two hundred acres willed to one "Zelda Selby" that caught the public unawares.[14]

Zelda Selby was identified in the will as Lucky Baldwin's daughter, and hasty investigation unearthed David Selby of Oakland, the husband of Zelda (or Rose Ella), who informed news reporters that his wife was the only issue of the "legal" marriage of Elias Baldwin and one Martha Agnes Fowler.[15] Records of the marriage apparently were never produced, but Baldwin's acknowledgment of his child was incontrovertible. Rose Ella Selby was awarded her bequest without objection, and the mystery of her parentage was left to speculation.

Serious objections to the will were raised instead by Lillie Bennett Baldwin. The widow was left "an undivided one-third interest" in the "Baldwin Annex" (adjacent to the Baldwin Hotel site in San Francisco),[16] an appraised inheritance of $160,000,[17] while her husband's daughters Clara and Anita were bequeathed "all the rest, residue, and remainder" of Lucky's real and personal property, said property "to be divided equally between them, share and share alike."[18]

The handling of Lillie Baldwin, et al. was inherited by Hiram Unruh, who was named executor of the estate by E.J., and by Bradner Lee, Lucky's attorney of many years, who not only had drawn up the will, but would defend it against its many challengers. Each man would be rewarded by law "one-half of one percent of the value of the estate, or $125,000."[19]

EJB, Anita, Zelda (Rose Ella), and Mistresses, ca. 1902

The appraised valuation of the real and personal property of Elias Baldwin, including some 33,500 remaining acres in Los Angeles County, was set by court appointees in 1910 at $10,919,801[20] but, as had been true for over a quarter of a century, Lucky Baldwin was deeply in debt.

At the time of death, most observers felt that there would be no estate for Unruh and Bradner Lee to either defend or wage battle over. Just a month and a half before he passed away E.J. "executed a Trust Deed to Hibernia Savings and Loan Society of San Francisco," reported Hull McClaughry, "for practically all of his property [the Ranchos La Puente, Potrero Grande, La Merced, and La Cienega] to secure a loan of $1,320,000." [21] The bank sought immediate repayment from the estate, said McClaughry, but was forestalled by Unruh and Attorney McNab, Anita Baldwin's counsel, and agreed instead to regular payments by the estate on both the principal and the interest due.[22]

Unruh, "under order of the court," was directed to "begin paying at between $25,000 and $27,000 a month," in addition to meeting "running expenses, widow's allowance, taxes, etc.," although, commented the *Los Angeles Examiner*, "all the actual money he could find was $4000."[23]

Unruh and Bradner Lee met with the heirs and with the consent of the court formulated a plan to subdivide and sell selected acreage to reduce the estate's indebtedness. The "Home Place," the most valuable of the Southland properties, was not then considered for sale, for E.J. had stipulated in his will that the land devolve, upon the deaths of Clara and Anita, to his daughters' children[24] "in equal shares, as tenants in common," with the "recommendation and request" that the homesite "be permanently kept and not disposed of."[25]

Rancho La Cienega was also reserved by Unruh, who contested the heirs' desire to sell the acreage for an offered $2,250,000. The property, Unruh argued, "would be worth $5 million more in five years' time."[26]

Executor Unruh, explained the *Examiner*, selected instead "the historic Potrero, the Felipe Lugo, and the Puente ranches and began cutting them up into ten- and twenty-acre tracts. He made these three old Spanish estates," said the newspaper, "save the Baldwin fortune."[27] The land sold well at $250 to $300 per acre, and within a year some four thousand acres had been marketed, much of it to "Eastern settlers."[28]

The cash infusion allowed Unruh to satisfy creditors, make mortgage payments, and meet fixed charges without threat. "Baldwin's enormous and world famous debts have been almost swept away," reported the *Examiner*. "Old Lucky's heirs, instead of losing their money, as was feared, will be richer than the old millionaire ever dreamed."[29] Said Unruh in 1910,

> I was with Mr. Baldwin thirty years, and in that time he paid to the Hibernia Savings Bank more than $4 million in interest alone. I do not think there has ever been another estate that has struggled under such a great burden of debt so long and finally come out solvent. People said after the death of Mr. Baldwin that we could never pull the estate out, and that besides we would be swamped with litigation. We have got past all the hard places. The claims are all provided for except the inheritance tax, and the cost of administration. A few little flies of suits are buzzing about but we will brush them out of the way when the time comes.[30]

The widow Baldwin, the newspapers reported, had engaged a firm of attorneys "for a tremendous onslaught against the estate," but

EJB and Clara, with Children

before suit could reach the courts, Clara and Anita agreed to a compromise settlement. Lillie Bennett Baldwin received "nearly a million dollars for herself and something in the neighborhood of half a million to her lawyers," it was said, payable from the first monies realized in Unruh's land sales.[31]

The settlement had been effected in hopes of circumventing additional litigation. "They knew that if the estate were ever carried into court and Baldwin's record laid bare," said the *Times*, "the scandal would attract any number of claimants, both fake and genuine."[32]

Their hopes proved to be in vain, though, for a number of Unruh's "little flies of suits" were filed against the estate. One of the first was initiated by Lillian Ashley Turnbull (the 1894 breach of promise suit), who sought a dividend for herself and for a child she claimed was Baldwin's.[33] Other would-be children and grandchildren filed similar suits, and a few contestants even challenged the identity of Elias Baldwin. A William Rutan, for instance, claimed that the real Elias J. Baldwin had been shot and killed on the wagon trip to California and that his own father had assumed E.J.'s name and identity.[34]

Bradner Lee met and defeated each of the fifty or so legal challenges to the Baldwin estate, though it took four years to complete the litigation.[35] The Mushet Audit Company, it was stated, required six volumes of "seventeen hundred large, closely-typewritten pages" to compile the estate reports,[36] but when the final settlement decree was issued April 24, 1913, the summation was heartening.

Cash, property, and rents valued at $20 million were turned over to Clara Baldwin Stocker and Anita Baldwin McClaughry, twice the appraised valuation of 1909, and beyond the $5 million already paid to meet debts, expenses, and administrative costs.[37] Further, $12 million worth of property had been sold during the course of administration, and "$1,200 per day," it was reported, was "coming to the estate in addition to the total valuation."[38]

Hiram Unruh and Bradner Lee each received $180,000 for their services and were publicly lauded by the bench. Said Judge Rives,

> I want to say as we finish with the Baldwin case, I think the management of the case should stand as an everlasting monument to the business sagacity of Mr. Unruh and his attorney Mr. Lee. I have watched them with interest from the commencement of it. They were beset with many kinds of difficulties—debts, law suits—pestered for four years. I think it is a most wonderful thing that at the expiration of this period they are now turning over an inheritance that is worth at least twice as much as when they took charge of it. Of course, that is very largely due to the wonderful development and growth of our glorious Southern California. But I have observed with a great deal of satisfaction and am able to state that this is one rich man's estate that has been carried through to such a splendid consummation, if you can call it that, as rapidly as this has. I do not know of any record that has been a better one, and I am glad that everybody is satisfied now and that the matter is at an end.[39]

Summary

Lucky Baldwin was a Southern California landowner, financier, rancher, horse breeder, horticulturist, conservationist, boomer, and city-maker. He was shrewd, energetic, acquisitive, amoral, sensitive to the beauties of nature, and indifferent to public opinion. "His was the spirit of a man of iron who fought for everything he wanted and never recognized a defeat," wrote the *Los Angeles Examiner*. "It will be long before the West sees another man like Elias Jackson Baldwin."[1]

E.J. was not a publicly charitable man. There are individuals who recount stories of personal generosity, but there were no Baldwin endowments, no hospitals, no schools, museums, or libraries with Baldwin wings or Baldwin collections. Lucky's societal contributions were indirect, and they were made during his lifetime, not after.

Lucky Baldwin's acquisition of Southern California ranch land in 1875 was celebrated at the time as a "herald of development."[2] "His enormous purchase in a section presumed to be a wilderness," reflected the *Daily Herald*, "awakened interest in Southern California to such a degree that the imitators of Baldwin were soon followed by an ever increasing army of land seekers."[3]

Although E.J. certainly did not single-handedly start the boom of the eighties, as is implied in the *Herald's* eulogy, he did play a significant role in the attendant development of the Southland. The towns of Sierra Madre and Monrovia grew directly out of Baldwin's land sales, and first "Baldwin," then Arcadia, were largely conceived in his image. Lucky Baldwin negotiated railroad contracts that linked both townsites and ranch lands to a transcontinental network, and he initiated building projects and improvements in Arcadia that not only gave employment to numbers of workmen, but also constituted material progress toward cityhood.

Baldwin's "work has, of course, been a selfish one," commented a *Times* reporter, "but he has opened up these thousands of acres to settlement, and his money has done what a number of smaller fortunes could not have done."[4]

Lucky Baldwin extolled the virtues of Southern California acreage, and at Santa Anita, he developed a showcase ranch that became a persuasive advertisement for future land sales. "My ranch is a benefit not only to California but to the whole United States," said Baldwin, "for anyone who will plant orchards and vineyards is a public benefactor."[5]

Baldwin Ranch produce was unsurpassed in quality and in quantity. The variety of output assured self-sustenance; cash crops supported a host of non-ranch activities; and specific products, notably Baldwin wines and brandies, were prize-winners with national recognition. The Baldwin racing stable was equally successful and lent credibility to E.J.'s contention that California bred and raised horses could seriously rival stock nurtured on Kentucky bluegrass.

Lucky Baldwin touted Southern California with the fervor of a convert, and he used his resources unhesitatingly to show his acreage to advantage. "Baldwin spent a million in making this ranch of his not only a splendidly productive property," wrote one observer, "but also one of the most beautiful estates ever laid out in this or any other country. It was his hobby, his pet."[6] The beauties of Santa Anita earned it acclaim as a tourist attraction and acknowledgment from Baldwin friends and critics alike that the creator, in the words of Bancroft, "must have other elements in his nature than those of the mere utilitarian and money-maker."[7]

Lucky Baldwin was in debt through most of his Southern California years, but "he was a man of great individuality, a positive man in every regard," noted the *Los Angeles Herald.* "When he decided to do anything, he set aside all thought of expense, trouble, hardship, and other obstacles and went ahead until he had accomplished fully all he had started out to do."[8]

Lucky Baldwin was in active pursuit of new goals until the day he died. "He always looked ahead," said a friend. "He always carried on."[9] The incorporation of Arcadia and the establishment of a racetrack "in his back yard" were the last and perhaps most flamboyant accomplishments of Baldwin's life, but they were typical of his Southern California boosterism. "I have made my money in California and I have spent it in California," said Baldwin. "This is the way, at least I think so, of advertising the state. Improve it and then the eastern people will find out for themselves whether it is a good place in which to live or not."[10]

The career of Lucky Baldwin was Southern California in microcosm from 1875 to 1909. His methods and motives of aggrandizement and promotion were his colorful own, but they both encouraged and reflected developments that metamorphised the whole of the Southland from the large-scale ranching of the late 1800s into the accelerating suburbanization in evidence by the turn of the century.

If Lucky were to return today he would find a network of cities occupying former ranch lands. The citrus, the vineyards, the fields of grain, the horses and livestock are gone. Only 127 acres of the huge Baldwin Ranch survived the lure of twentieth century subdivision, but it is the heart of the ranch, the homesite at the lake, that has been preserved within today's Los Angeles State and County Arboretum. E.J. would surely be pleased that tourists still admire his cottage and matching coach barn and still rest in the shade of trees he introduced to the Southern California landscape.

"He was Lucky Baldwin, perhaps the outstanding individualist of his generation," concluded his biographer. "His was the spirit of the frontier."[11]

REFERENCES

Introduction

1. Ralph D. Paine, "Where Ranch and City Meet," *Outing*, XLVIII, (August 1906), 539.
2. Karen Saunders, "The Story of Lucky Baldwin" (term paper, University of California, Los Angeles, Extension, March 16, 1970), p. 13.
3. Lowry McCaslin, personal interview, Arcadia, California, July 2, 1979.
4. Saunders, p. 13.
5. C.B. Glasscock, *Lucky Baldwin: The Story of an Unconventional Success* (Indianapolis: Bobbs-Merrill, 1933), pp. 16-26. H.H. Bancroft states that Baldwin was born in Butler County, Ohio, near the town of Hamilton.
6. "Baldwin, Elias Jackson," *The National Cyclopedia of American Biography* (1932), XXII, 382.
7. Glasscock, pp. 44-46.
8. Glasscock, p. 47.
9. Glasscock, pp. 49-66.
10. Glasscock, pp. 67-73.
11. Glasscock, pp. 81-85.
12. Hubert Howe Bancroft, *Chronicles of the Builders of the Commonwealth*, III (San Francisco: The History Company, 1892), p. 337
13. Bancroft, III, p. 338.
14. Glasscock, pp. 92-113.
15. Bancroft, III, p. 341.
16. Bancroft, III, p. 344.
17. Glasscock, p, 119.
18. Glasscock, pp. 134-39.
19. Glasscock, p. 139.
20. Glasscock, p. 141.
21. Glasscock, pp. 151-54.
22. Glasscock, pp. 155-56.
23. *Pacific Coast Annual Mining Review and Stock Ledger* (San Francisco: Frances and Valentine, 1878), p. 70.
24. George Lyman, *Ralston's Ring: California Plunders the Comstock* (New York: Charles Scribner's Sons, 1937), p. 258.

CHAPTER ONE

Land Acquisition and Initial Development

1. *Los Angeles Evening Express*, "Local Items," March 20, 1875, p. 3.
2. Ibid.
3. John W. Robinson, "A Mountain of Gold," *Desert*, XXXVII, (September 1975), 35-36.
4. *Los Angeles Evening Express*, "What Lucky Baldwin Thinks," February 15, 1875, p. 2.
5. Hubert Howe Bancroft, *Chronicles of the Builders of the Commonwealth*, III (San Francisco: The History Company, 1892), p. 361.
6. *Express*, "What Lucky Baldwin Thinks," February 15, 1875, p. 2.
7. *Los Angeles Evening Express*, "Gold Mountain," March 19, 1875, p. 3.
8. Robinson, pp. 36-37.
9. *Los Angeles Times*, "Lucky Baldwin Proud Monarch of Queerest of Little Kingdoms," November 3, 1907, p. 1.
10. Ibid.
11. Maurice H. Newmark, and Marco R. Newmark, eds., *Sixty Years in Southern California, 1853-1913. Containing the Reminiscences of Harris Newmark* (3d ed.; Boston and New York: Houghton Mifflin, 1930), p. 474.
12. *Los Angeles Evening Express*, "Splendid Summer House," March 20, 1875, p. 3.
13. For reasons unexplained, though perhaps tax-related, the deed of March 20 conveyed the Santa Anita property from Harris Newmark to Thomas B. Fawcett, Baldwin's brother-in-law. A second deed, filed April 8, 1875, reconveyed the property, for a consideration of $5, from T.B. Fawcett to E.J. Baldwin (Los Angeles County, Hall of Records, *Deeds*, Book 40, pp.212-14).
14. Los Angeles County, Hall of Records, *Deeds*, Book 34, pp. 253-56.
15. Ibid. All of the water of Little Santa Anita Canyon plus half the water of Sawpit Canyon had previously accrued to Rancho Santa Anita by virtue of long use.
16. *Los Angeles Evening Express*, "Most Important Transaction of the Day," March 22, 1875, p. 2.
17. *Los Angeles Daily Herald*, "Santa Anita," March 21, 1875, p. 2.
18. *Los Angeles Evening Express*, "How It Works," March 25, 1875, p. 3.
19. *Los Angeles Times*, "Millions the Stake Here," November 28, 1912, Pt. II, p. 3.
20. Los Angeles County, Hall of Records, *Deeds*, Book 39, pp. 404-406.
21. Los Angeles County, Hall of Records, *Miscellaneous Records*, Book 42, "Map of the Western Two-Thirds of Rancho San Francisquito," pp. 93-94.
22. C.B. Glasscock, *Lucky Baldwin: The Story of an Unconventional Success* (Indianapolis: Bobbs-Merrill, 1933), p. 176.
23. California Scrapbook-Biography (collection of undated newspaper clippings), "Woman is Loser in $2,600,000 Estate Suit," California Room, Los Angeles Public Library.
24. *Los Angeles Daily Herald*, "Lucky Baldwin and His Ranch," February 6, 1876, p. 3.
25. *Express*, "How It Works," March 25, 1875, p. 3.
26. *Pacific Coast Annual Mining Review and Stock Ledger* (Frances and Valentine, 1878), pp. 71-72. A section of biographical sketches includes a four-page article on

Elias Baldwin with mention of his current Southern California activities.
27. *Los Angeles Evening Express*, "The Santa Anita Ranch," October 14, 1876, p. 3.
28. *Pacific Coast Annual Mining Review*, pp. 71-72.
29. Ibid, p. 72.
30. *Herald*, "Lucky Baldwin and His Ranch," February 6, 1876, p. 3.
31. Remi Nadeau, *City-Makers: The Men Who Transformed Los Angeles From Village to Metropolis During the First Great Boom, 1868-78* (Garden City: Doubleday, 1948), p. 50.
32. Nadeau, pp. 213-16.
33. *Herald*, "Lucky Baldwin and His Ranch," February 6, 1876, p. 3.
34. Los Angeles County, Hall of Records, *Mortgages*, Book 17, pp. 50-55.
35. Ibid.
36. Mr. and Mrs. Temple and Juan Sanchez sold 203 acres to E.J. Baldwin, January 7, 1876, for $10,150 (Los Angeles County, Hall of Records, *Deeds*, Book 42, pp. 74-76).
37. An amendment excepted 297 acres sold by Temple and Workman to E.J. Baldwin for $10,000 (Los Angeles County, Hall of Records, *Deeds*, Book, 40 pp. 233-35).
38. Abstract of Title and Certificate of Search to Proceedings in the Case of Camilo Martin vs. William Temple et al., April 21, 1888, Anita M. Baldwin Collection, Box 6, Rare Book Room, Huntington Library. This case served as the vehicle through which the order for sale of the mortgaged Temple/Workman/Sanchez lands was filed.
39. Ibid.
40. Los Angeles County, Hall of Records, *Deeds*, Book 40, pp. 232-33.
41. Los Angeles County, Hall of Records, *Deeds*, Book 173, pp. 493-95.
42. Nadeau, p. 219.
43. Abstract of Title, Camilo Martin vs. William Temple.
44. Nadeau, pp. 219-22.
45. Abstract of Title, Camilo Martin vs. William Temple.
46. Ibid.
47. For reasons unascertained, Camilo Martin, of San Francisco, was assigned the promissory notes by E.J. Baldwin, and it was in Martin's name that the decree of sale was ordered. The property was later conveyed to Baldwin at a nominal price (Bradner W. Lee Papers, Box 3, Special Collections, Research Library, University of California, Los Angeles).
48. Abstract of Title, Camilo Martin vs. William Temple.
49. Ibid.
50. Los Angeles County, Hall of Records, *Deeds*, Book 77, pp. 162-63.
51. Los Angeles County, Hall of Records, *Deeds*, Book 76, pp. 288, 304-306.
52. Judson A. Grenier, ed., *A Guide to Historic Places in Los Angeles County* (Dubuque: Kendall/Hunt Publishing Co., 1978), p. 225.
53. Los Angeles County, Hall of Records, *Deeds*, Book 38, pp. 229-30.
54. Estate Papers of Elias Jackson Baldwin, "Report to Superior Court, County of Los Angeles, by Hiram A. Unruh, Executor," Report 1, April 9, 1909 to March 31, 1910, Anita M. Baldwin Collection, Box 6, Rare Book Room, Huntington Library.
55. *Los Angeles Tribune*, "From an Interview in the *Louisville Courier-Journal*," May 23, 1887, p. 6.
56. Bancroft, III, p. 353.

57. "California's Choicest Locality—San Gabriel Valley" (brochure), 1891-92, Local History Collection, Arcadia Public Library.
58. Walter Lindley and J.P. Widney, *California of the South* (New York: D. Appleton, 1888), p. 163.
59. "Inventory and Appraisement in the Matter of the Estate of Elias J. Baldwin," June 13, 1910, Anita M. Baldwin Collection, Box 6, Rare Book Room, Huntington Library.
60. Statement by Samuel Butler, an early ranch employee, interview by Elizabeth Dixon, Oral History Program, University of California, Los Angeles, 1966. (Transcript.)
61. Ralph D. Paine, "Where Ranch and City Meet," *Outing*, XLVIII (August, 1906), 532.
62. *Los Angeles Examiner*, "$20 Million for Two Baldwin Heirs," April 2, 1912, p. 2.

CHAPTER TWO

The Boom of the Eighties

1. Hubert Howe Bancroft, *Chronicles of the Builders of the Commonwealth*, III (San Francisco: The History Company, 1892), p. 350.
2. C.B. Glasscock, *Lucky Baldwin: The Story of an Unconventional Success* (Indianapolis: Bobbs-Merrill, 1933), p. 203.
3. Anita Baldwin, Birth Certificate, January 10, 1876, Anita M. Baldwin Collection, Box 4, Rare Book Room, Huntington Library.
4. Glasscock, pp. 178-79.
5. Ibid, p. 218.
6. Although the exact date of marriage is uncertain, both the *San Francisco Daily Alta* of March 11, 1879 (cited in Glasscock, p. 207) and the *Los Angeles Daily Herald* of March 16, 1879 (p. 3), carry references to E.J. Baldwin and his "bride," Jennie Dexter.
7. Glasscock, p. 219.
8. *Los Angeles Examiner*, "$20 Million for Two Baldwin Heirs," April 2, 1912, p. 2.
9. *Los Angeles Times*, "Busy Life Is Snuffed Out," December 17, 1916, Pt. II, p. 3.
10. Ibid.
11. *Examiner*, "$20 Million for Two Baldwin Heirs," April 2, 1912, p. 2.
12. *Times*, "Busy Life Is Snuffed Out," December 17, 1916, Pt. II, p. 3.
13. Glasscock, p. 221.
14. *Los Angeles Evening Express*, "How It Works," March 25, 1875, p. 3.
15. *Los Angeles Daily Herald*, "Local Brevities," October 14, 1875, p. 3.
16. Harold Carew, *History of Pasadena and the San Gabriel Valley* (Los Angeles: S.J. Clarke Publishing Company, 1930), p. 363. Beginning in 1874 and continuing for the next twenty-five years, Nathaniel Carter made annual "excursions" to his Massachusetts home to praise the Southern California climate and entice easterners to view the Southland first-hand.
17. Los Angeles County, Hall of Records, *Deeds*, Book 78, pp. 545-48.
18. *Sierra Madre Vista*, "Sierre Madre Tract," July, 1881, p. 8.

19. Carew, p. 367.
20. *Sierra Madre Vista*, "Vineyard, Orchard, and Farming Lands for Sale in the Santa Anita Tract" (map and advertisement), October, 1883, p. 4. Plans for the second tract, originally called "Santa Anita Colony," were mentioned as early as 1881 (*Sierra Madre Vista*, "Sierra Madre Tract," July, 1881, p. 8).
21. *Sierra Madre Vista*, "Santa Anita Ranch," July, 1881, p. 5.
22. *Sierra Madre Vista*, "Vineyard, Orchard, and Farming Lands for Sale" (map and advertisement), October, 1883, p. 4.
23. Ibid.
24. Louis Wolfskill inherited Rancho Santa Anita from his father, pioneer William Wolfskill, in 1865. In1872 he sold the remaining eight thousand acres of Santa Anita to Harris Newmark.
25. *Monrovia Messenger*, "An Explanation," May 2, 1889, p. 1.
26. John Wiley, *History of Monrovia* (Pasadena: Pasadena Star-News, 1927), p. 44.
27. *Sierra Madre Vista*, "Sierra Madre and Santa Anita Tracts," October, 1883, p. 4.
28. "By-Laws of Santa Anita Water Company, Los Angeles County, California—Santa Anita Tract," Rare Book Room, Huntington Library.
29. Seth Colver, ed., *The Historical Volume and Reference Works: Covering Arcadia, Azusa, Baldwin Park, Bradbury, Charter Oak, Claremont, Covina, Duarte, El Monte, Glendora, Irwindale, La Verne, Monrovia, Pomona, San Dimas, South El Monte, West Covina* (Arlington, California: Historical Publishers, 1964), p. 273.
30. W.W. Robinson, *Ranchos Become Cities* (Pasadena: San Pasqual Press, 1939), pp. 194-95.
31. Robinson, p. 195.
32. Glenn S. Dumke, *The Boom of the Eighties in Southern California* (San Marino: Huntington Library, 1944), p. 79.
33. Joseph Netz, "The Great Los Angeles Real Estate Boom of 1887," *Annual Publications—Historical Society of Southern California*, X (1915-16), 57.
34. Los Angeles and San Gabriel Valley Rail Road Company, Contract with E.J. Baldwin, November 24, 1885, Historical Archives, Los Angeles State and County Arboretum. (Photocopy.)
35. The Los Angeles and San Gabriel Valley Railway, "Prospectus," circa 1884, Rare Book Room, Huntington Library.
36. Dr. Hiram A. Reid, *History of Pasadena* (Pasadena: Pasadena History Company, Publishers, 1895), pp. 427-29.
37. Los Angeles and San Gabriel Valley Railway, "Prospectus," circa 1884.
38. Los Angeles and San Gabriel Valley Rail Road Company, Contract with E.J. Baldwin, November 24, 1885.
39. Ibid.
40. Ibid.
41. *Los Angeles Daily Herald*, "The Santa Anita Tract on Sale in Five to Thirty Acre Lots" (advertisement), January 6, 1886, p. 5.
42. "Map of the Town of Baldwin in the Santa Anita Tract, May 1886," Rare Book Room, Huntington Library.

43. *Herald*, "The Santa Anita Tract on Sale" (advertisement), January 6, 1886, p. 5.
44. Ibid.
45. *Los Angeles Daily Herald*, "The Work of Development in the Foothills," June 1, 1886, p. 2.
46. Ibid.
47. Ibid.
48. Los Angeles County, Hall of Records, *Deeds*, Book 150, pp. 103-105.
49. *Los Angeles Times*, "A New Colony—Monrovia," January 1, 1886, p. 5.
50. Wiley, p. 51
51. *Los Angeles Daily Herald*, "Monrovia Movements," December 9, 1886, p. 6.
52. Wiley, p. 63.
53. Los Angeles County, Hall of Records, *Deeds*, Book 151, pp. 547-49. Unruh's name appears as grantor in several later Santa Anita Tract deeds, but at the time of Baldwin's death in 1909, all of the unsold Santa Anita Tract acreage was inventoried as property of the deceased ("Inventory and Appraisement in the Matter of the Estate of Elias J. Baldwin," June 13, 1910).
54. "Map of a Part of Santa Anita Tract, September, 1886," Rare Book Room, Huntington Library.
55. *Los Angeles Daily Herald* , "Monrovia Station—Santa Anita Tract" (advertisement), January 23, 1887, p. 2.
56. "Map of a Part of Santa Anita Tract, September, 1886."
57. *Los Angeles Daily Herald*, "Santa Anita Tract—Part of the Celebrated Santa Anita Ranch, Now on Sale" (advertisement), January 23, 1887, p. 2.
58. *Monrovia Planet*, "A Local Railroad—Orange Belt Route Through the San Gabriel Valley," January 8, 1887, p. 2.
59. *Los Angeles Times*, "Railways—The Three Best," May 5, 1887, p. 2.
60. Letter from Thomas Fitzgerald, General Attorney, California, for The Atchison, Topeka, and Santa Fe Railway Company, July 2, 1979. The California Central was further consolidated on November 7, 1889, into the Southern California Railway Company, and on January 17,1906, Southern California Railway was sold to The Atchison, Topeka, and Santa Fe (Letter from Thomas Fitzgerald).
61. Cary McWilliams, *Southern California: An Island on the Land* (Santa Barbara: Peregrine Smith, 1946), p. 118.
62. On one well-publicized day, the Southern Pacific-Santa Fe rate war produced a $1 transcontinental passenger fare (McWilliams, p. 118).
63. *Monrovia Planet*, "Monrovia's Future Railroads," March 19, 1887, p. 2.
64. *Los Angeles Times*, "The Santa Fe's Enormous Deal Near the Raymond," April 7, 1887, p. 1.
65. C.F. Shoop, "History of City's First Railroad Line Traced" (newspaper clipping), Vertical File, Los Angeles State and County Arboretum.
66. *Monrovia Planet*, "The New Railway," January 22, 1887, p. 2. The advertising map seen in Figure 3 depicts a "Monrovia, Arcadia, Pasadena R.R." just north of the California Central Railway. This nonexistent line was probably either the proposed Santa Fe cut-off or, more likely, the Southern Pacific branch survey ("Birdseye View of Arcadia and Santa Anita Tract," circa 1887).
67. *Los Angeles Daily Herald*, "Beautiful Arcadia," March 30, 1887, p. 1.

68. *Los Angeles Daily Herald*, "The New Railroad," January 18, 1887, p. 8.
69. *Monrovia Planet*, "A New Town," March 12, 1887, p. 2.
70. *Monrovia Planet*, "Satellites," March 26, 1887, p. 3.
71. *Herald*, "Beautiful Arcadia," March 30, 1887, p. 1.
72. *Los Angeles Daily Herald*, "Arcadia" (advertisements), April 1 to June 9, 1887.
73. Ibid.
74. *Monrovia Planet*, "Arcadia—Monrovia's Youngest Sister and Her Progress," April 9, 1887, p. 2.
75. *Monrovia Planet*, "Satellites," April 23, 1887, p. 3.
76. *Monrovia Planet*, "Arcadia—Monrovia's Youngest Sister," April 9, 1887, p. 2.
77. *Los Angeles Daily Herald*, "News Notes," April 3, 1887, p. 10.
78. *Monrovia Planet*, "Santa Anita Tract" (advertisement), June 11, 1887, p. 6.
79. *Los Angeles Daily Herald*, "Arcadia," May 21, 1887, p. 5.
80. *Monrovia Planet*, "A Perfume Factory," March 5, 1887, p. 4.
81. *Monrovia Planet*, "Satellites," April 9, 1887, p. 1.
82. *Monrovia Planet*, "Things of Beauty—A Visit to Arcadia's Beautiful Flower Farm," June 4, 1887, p. 3.
83. Ibid.
84. *Los Angeles Daily Herald*, "Local News Notes," March 23, 1886, p. 6.
85. *Monrovia Planet*, "Arcadia—Monrovia's Youngest Sister," April 9, 1887, p. 2.
86. "Map of a Part of Arcadia, 1887," Rare Book Room, Huntington Library.
87. *Monrovia Planet*, "A Monster Hotel," April 2, 1887, p. 2.
88. Ibid.
89. *Los Angeles Daily Herald*, "Beautiful Arcadia," March 30, 1887, p. 1.
90. *Planet*, "A Monster Hotel," April 2, 1887, p. 2.
91. *Los Angeles Daily Herald*, "Arcadia," May 5, 1887, p. 12.
92. *Los Angeles Daily Herald*, "News Notes," June 7, 1887, p. 10.
93. *Monrovia Messenger*, "The Hotel Oakwood: Arcadia's Pretty Hostelry Open for the Winter," December 26, 1889, p. 4.
94. Ibid.
95. *Los Angeles Times*, "Incorporated," May 8, 1887, p. 2.
96. *Los Angeles Daily Herald*, "Rapid Transit," July 10, 1887, p. 7.
97. *Los Angeles Daily Herald*, "L.A. and S.G.V.R.T.R.R.," July 21, 1887, p. 11.
98. Franklyn Hoyt, "The San Gabriel Valley Rapid Transit Railroad," *Southern California Quarterly*, XXXIII (September, 1951), 215-16.
99. Ibid.
100. Robert Studer, ed., *The Historical Volume and Reference Works: Including Alhambra, Monterey Park, Rosemead, San Gabriel, South San Gabriel, Temple City* (Los Angeles: Historical Publishers, 1962), p. 82.
101. "Pocket Time Table—California Central Ry. and California Southern R.R." (brochure), October 7, 1888, Rare Book Room, Huntington Library.
102. Seth Colver, ed., *Historical Volume and Reference Works: Covering Arcadia, Azusa, Baldwin Park, Bradbury, Charter Oak, Claremont, Covina, Duarte, El Monte, Glendora, Irwindale, La Verne, Monrovia, Pomona, San Dimas, South El Monte, West Covina*

(Arlington, California: Historical Publishers, 1964), p. 31.
103. *Los Angeles Daily Herald*, "Felipe Lugo Tract" (advertisement), January 23, 1887, p. 11; "Map Showing Subdivisions of the Rancho Potrero de Felipe Lugo, September, 1886," Rare Book Room, Huntington Library.
104. "California's Choicest Locality—San Gabriel Valley" (brochure), 1891-92, Local History Collection, Arcadia Public Library.
105. "How to Get a Deed from E.J. Baldwin—Lands in the San Gabriel Valley Offered for Sale by E.J. Baldwin" (brochure), 1891-92, Local History Collection, Arcadia Public Library.
106. Los Angeles County, Hall of Records, *Miscellaneous Records*, Book 42, "Map of Santa Anita Colony," p. 87.
107. "How to Get a Deed" (brochure), 1891-92.
108. Ibid.
109. Los Angeles County, Hall of Records, *Miscellaneous Records*, Book 52, "E.J. Baldwin's Addition No. 1 to Santa Anita Colony," p. 60; Los Angeles County, Hall of Records, *Miscellaneous Records*, Book 53, "E.J. Baldwin's Addition No. 2 to Santa Anita Colony," p. 4.
110. "How to Get a Deed" (brochure), 1891-92.
111. "California's Choicest Locality" (brochure), 1891-92.
112. Ibid.

CHAPTER THREE

The Ranch—Produce, Pride, and Profits

1. "California's Choicest Locality—San Gabriel Valley" (brochure), 1891-92, Local History Collection, Arcadia Public Library.
2. *Pacific Coast Annual Mining Review and Stock Ledger* (San Francisco: Frances and Valentine, 1878), p. 72.
3. Hubert Howe Bancroft, *Chronicles of the Builders of the Commonwealth*, III (San Francisco: The History Company, 1892), p. 359.
4. Elizabeth Cleminson, "Memory of E.J. Baldwin Ranch, 1889-1890" (hand-drawn map), Historical Archives, Los Angeles State and County Arboretum.
5. Fraser, "E.J. Baldwin—Notes Taken During a Visit to His Santa Anita Rancho" (dictations and materials for H.H. Bancroft's *Chronicles of the Builders of the Commonwealth*), August 26, 1891, Bancroft Library, University of California, Berkeley. (Transcript.)
6. *Los Angeles Daily Herald*, "Santa Anita Station," June 13, 1886, p. 5.
7. *Sierra Madre Vista*, "Local News and Notes," March 24, 1888, p. 3.
8. *Sierra Madre Vista*, "Local News and Notes," April 14, 1888, p. 3.
9. *Sierra Madre Vista*, "About a Depot Building," March 9, 1889, p. 2.
10. Ibid.
11. The Santa Anita Depot was removed from the path of the Foothill (210) Freeway in 1968 and reconstructed a half mile south, on the grounds of the Los Angeles State

and County Arboretum.
12. *Sierra Madre Vista*, "Local Views," August 21, 1890, p. 3.
13. Ibid.
14. *Sierra Madre Vista*, "Santa Anita and Mr. Baldwin," November 6, 1890, p. 3.
15. *Sierra Madre Vista*, "Local Views," August 28, 1890, p. 3.
16. *Sierra Madre Vista*, "Local Views," November 13, 1890, p. 3.
17. Ibid.
18. Ibid. It has not been ascertained whether the privilege was a contractual or merely a verbal agreement.
19. *Sierra Madre Vista*, "Railroad Time Table," May 21, 1891, p. 3. Glasscock relates a since popularized story (apparently given to him by ranch employee Charlie Anderson) of a Baldwin confrontation with Santa Fe over E.J.'s right to disembark on signal from all Santa Fe trains. Returning from Bear Valley, the story goes, E.J. was denied a ticket to Santa Anita on the overland train then coming through. Baldwin ordered a telegram blank and wrote: "J.F. Falvey, Superintendent, Santa Anita Ranch. Put two hundred men to work at once tearing up Santa Fe tracks through my ranch." E.J. received his ticket at once (C.B. Glasscock, *Lucky Baldwin: The Story of an Unconventional Success* (Indianapolis: Bobbs-Merrill, 1933), p. 250.
20. Fraser.
21. Ibid.
22. *Sierra Madre Vista*, "Local Views," July 3, 1890, p. 3.
23. *Sierra Madre Vista*, "Santa Anita Store" (advertisement), July 3, 1890, p. 1.
24. Julian Fisher, personal interview, Monrovia, California, November 2, 1976.
25. *Los Angeles Daily Herald*, "Local News Notes," January 17, 1886, p. 6.
26. Fisher, personal interview, November 2, 1976.
27. Ibid.
28. Samuel Butler, interview by Elizabeth Dixon, Oral History Program, University of California, Los Angeles, 1964. (Transcript.)
29. Julian Fisher, personal interview, Monrovia, California, August 4, 1976.
30. Bancroft, III, p. 370.
31. Elizabeth Cleminson, interview by Barbara Fincher, Temple City, California, October, 1976.
32. Glasscock, p. 241.
33. Fraser.
34. Bancroft, III, p. 356.
35. Henry Stull, "A Day at the 'Home of Grinstead,'" *Spirit of the Times*, CXXXVII (March 18, 1899), 862.
36. Bancroft, III, p. 356.
37. Ibid.
38. B.K. Beckwith, *The Story of Santa Anita—The Great Race Place* (Los Angeles: Day Printing Corp., 1975), p. 8.
39. Glasscock, p. 238.
40. Carleton Burke, "Pastime of Millions," *The Thoroughbred*, IV (April-May, 1945), 32.
41. *Los Angeles Times*, "Lucky Baldwin—A Look at His Famous Ranch and Its Wonders,"

June 11, 1893, p. 9.

42. Bancroft, III, p. 356.
43. Burke, p. 17.
44. Glasscock, p. 239.
45. Stull, p. 862.
46. *Monrovia Planet*, "Baldwin's Stables," April 30, 1887, p. 1.
47. Beckwith, p. 7.
48. Ibid.
49. Beckwith, p. 8.
50. Ibid.
51. Los Angeles Racing Association, *Souvenir, Santa Anita Park, Arcadia, Cal. Season 1908 and 1909* (Los Angeles: J.C. Effinger, 1908), California Turf Association Library, Arcadia, California.
52. Beckwith, p 8.
53. Bancroft, III, p. 356.
54. *Los Angeles Times*, "Lucky Baldwin," January 28, 1895, p. 3.
55. Fraser.
56. "California's Choicest Locality" (brochure), 1891-92.
57. *Sierra Madre Vista*, "Local Views," August 27, 1891, p. 3.
58. Fraser.
59. Bancroft, III, p. 354.
60. *Sierra Madre Vista*, "Notes and News," January 15, 1891, p. 2.
61. Bancroft, III, p. 354.
62. *Los Angeles Times*, "Lucky Baldwin's Latest Fortune," March 9, 1902, Pt. IV, p. 1.
63. Bancroft, III, p. 354.
64. Mechanical, dam-like "partitioners" separated Baldwin's water from that of the Santa Anita and Sierra Madre Water Companies, respectively.
65. William Hammond Hall, *Irrigation in Southern California* (Sacramento: State Printing Office, 1888), pp. 460-64.
66. Hall, p. 476.
67. *Los Angeles Times*, "Lucky Baldwin—A Look at His Famous Ranch and Its Wonders," June 11, 1893, p. 9.
68. *Sierra Madre Vista*, "Local News and Notes," August 2, 1889, p. 8.
69. Cleminson, interview by Barbara Fincher, October, 1976.
70. "Prince of the House of Argonauts," *San Gabriel Valley Digest*, III (April, 1930), 6.
71. Fisher, personal interview, August 4, 1976.
72. Fraser.
73. Ibid.
74. *Monrovia Messenger*, "Local News," November 14, 1889, p. 4; Bancroft, III, p. 358.
75. Bancroft, III, p. 358.
76. *Los Angeles Times*, "Lucky Baldwin—A Look at His Famous Ranch and Its Wonders," June 11, 1893, p. 9.
77. The name "Queen Anne Cottage" was not applied to the house until after Baldwin's death.

Its origin is obscure, though the reference is probably to the style of architecture evidenced in the octagonal tower and wrap-around verandah.

78. "Sierra Madre and Santa Anita," *Pacific Rural Press*, May 9, 1885 (magazine clipping), Historical Archives, Los Angeles State and County Arboretum.
79. *Los Angeles Times*, "Baldwin's Belvedere," May 8, 1886, p. 1.
80. Ibid. The Queen Anne Cottage was designated State Historic Landmark No. 367 in 1940 in response to an application filed by Wesley Davies, representative of Rancho Santa Anita Inc. (then owners of the homesite). In the application, A.A. Bennett was cited by Mr. Davies as the architect of the Cottage, and the building completion date was listed as 1881. No evidence has been found by this writer to either support or refute Bennett's role in the design of the Cottage and, as previously stated, construction appears to have been completed in 1885-86.
81. Fraser.
82. *Los Angeles Times*, "Lucky Baldwin—Proud Monarch of Queerest of Little Kingdoms," November 13, 1907, Pt. II, p. 1.
83. "Inventory and Appraisement in the Matter of the Estate of Elias J. Baldwin," June 13, 1910, Anita M. Baldwin Collection, Box 6, Rare Book Room, Huntington Library.
84. Bancroft, III, p. 355.
85. Ibid.
86. "Baldwin's Ranch. Winery and Price List. M. Lawrence and Co., Sole Agents" (pamphlet), Research Library, Los Angeles State and County Arboretum.
87. Frona Wait, *Wines and Vines of California: A Treatise on the Ethics of Wine Drinking* (San Francisco: The Bancroft Company, 1889), pp. 179-80.
88. Fraser. The brick sherry building is the only winery structure standing today. It is located near the head of the turf course at Santa Anita racetrack and is used by park management as a storage shed.
89. Bancroft, III, p. 355.
90. Letter from Ray Knisley, former Baldwin Ranch manager, to Jack McCaskill, Los Angeles State and County Arboretum, January, 1968.
91. "Baldwin's Ranch. Winery and Price List" (pamphlet).
92. *Los Angeles Times*, "Lucky Baldwin—A Look at His Famous Ranch and Its Wonders," June 11, 1893, p. 9.
93. *The Illustrated Los Angeles Herald*, "A Rich Possession—The Home of E.J. Baldwin, Esq., Los Angeles County," June, 1889, p. 21.
94. "California's Choicest Locality" (brochure), 1891-92.
95. Claire Radford, *La Puente Kaleidoscope* (La Puente: La Puente Valley Women's Club, 1974), pp. 12-13. The warehouse was located at the intersection of today's Valley Boulevard and Glendora Avenue.
96. Donald L. Hathcock, "The History of Baldwin Park, California" (Master's thesis, Whittier College, 1954), p. 68.
97. Seth Colver, ed., *The Historical Volume and Reference Works: Covering Arcadia, Azusa, Baldwin Park, Bradbury, Charter Oak, Claremont, Covina, Duarte, El Monte, Glendora, Irwindale, La Verne, Monrovia, Pomona, San Dimas, South El Monte,*

West Covina (Arlington, California: Historical Publishers, 1964), p. 357.
98. Cleminson, interview by Barbara Fincher, October, 1976.
99. *San Gabriel Valley News*, "El Monte," April 11, 1878, p. 2.
100. Fisher, personal interview, November 2, 1976.
101. *Los Angeles Evening Express*, "Local Dots," May 6, 1877, p. 3.
102. Cleminson, interview by Barbara Fincher, October, 1976.
103. Ibid.
104. Glasscock, p. 243.

CHAPTER FOUR

Land Poor and Short On Luck

1. *Los Angeles Times*, "Baldwin's Big Mortgage," January 24, 1896, Pt. I, p. 6.
2. H.A. Unruh, letter to C.A. Canfield, August 17, 1908, Anita M. Baldwin Collection, Box 5, Rare Book Room, Huntington Library.
3. James M. Guinn, *A History of California and an Extended History of Its Southern Coast Counties*, I (Los Angeles: Historic Record Company, 1907), p. 306.
4. *Los Angeles Times*, "Baldwin's Big Mortgage," January 24, 1896, Pt. I, p. 6.
5. *Los Angeles Times*, "Smoking Ruin—Baldwin Hotel Entirely Burned Down," November 24, 1898, Pt. I, p. 4.
6. *Times*, "Baldwin's Big Mortgage," January 24, 1896, Pt. I, p. 6.
7. Jackson Graves, *California Memories, 1857-1930* (Los Angeles: Times Mirror Press, 1930), p. 11.
8. H.A. Unruh, letter to C.A. Canfield, August 20, 1908, Anita M. Baldwin Collection, Box 5, Rare Book Room, Huntington Library.
9. C.B. Glasscock, *Lucky Baldwin: The Story of an Unconventional Success* (Indianapolis: Bobbs-Merrill, 1933), p. 205.
10. Glasscock, pp. 209-10.
11. Hubert Howe Bancroft, *Chronicles of the Builders of the Commonwealth*, III (San Francisco: The History Company, 1892), p. 350.
12. *Sierra Madre Vista*, "The Baldwin," October, 1883, p. 5.
13. Glasscock, pp. 192-93.
14. Bancroft, III, p. 352.
15. Glasscock, pp. 188, 221.
16. Oscar Shuck, *Historical Abstract of San Francisco,* I (San Francisco: Oscar Shuck, 1897), p. 17.
17. Edward B. Scott, *The Saga of Lake Tahoe* (Lake Tahoe: Sierra-Tahoe Publishing Company, 1957), p. 152.
18. Scott, pp. 156-59.
19. Scott, pp. 159-61.
20. Scott, pp. 160-63.
21. Mushet Audit Company and Bradner W. Lee, "In the Matter of the Estate of Elias J. Baldwin, Deceased. Inventory of Books, Documents, Maps, Correspondence, Miscellaneous

Papers, Etc.," April 16, 1913, Anita M. Baldwin Collection, Volume I, Rare Book Room, Huntington Library.
22. *Sierra Madre Vista*, "Local Views," October 2, 1890, p. 3.
23. *Los Angeles Times*, "E.J. Baldwin's Vast Irrigating Enterprise," August 20, 1891, p. 7.
24. *Sierra Madre Vista*, "Local Views," July 23, 1891, p. 3.
25. *Sierra Madre Vista*, "Local Views," May 28, 1891, p. 3.
26. Mushet Audit Company and Bradner W. Lee.
27. John W. Robinson, "A Mountain of Gold," *Desert*, XXXVIII (September, 1975), 37.
28. "E.J. Baldwin Account With Lloyd and Wood, Attorneys at Law, 1896-97," Anita M. Baldwin Collection, Box 5, Rare Book Room, Huntington Library. Baldwin was billed by his attorneys for drawing a deed of sale of land to Budd Doble and Ira Taylor (1896) and for work "in relation to selling Gold Mountain Mine to parties in New York" (1897). In 1908, Hiram Unruh estimated remaining Baldwin holdings in Bear Valley as "twenty-seven hundred acres of mountain and timber lands" (Unruh, letter to Canfield, August 10, 1908).
29. Glasscock, p. 227. Lillie Bennett was the daughter of architect A.A. Bennett.
30. *Pasadena Star News*, "Building of Original Ranch Units Remembered" (clipping), June 10, 1956.
31. *Los Angeles Times*, "His Girl Wife Had Patience," March 2, 1909, Pt. I, p. 1.
32. Ibid.
33. Ibid.
34. Glasscock, p. 152.
35. Glasscock, p. 153.
36. Judge Donald William Hamblin, "Lucky's Litigious Loves," *The Branding Iron*, March, 1969, p. 5.
37. Glasscock, pp. 229-30.
38. *Los Angeles Times*, "Baldwin vs. Bell," January 15, 1886, p. 4.
39. Ibid; *Los Angeles Times*, "Angry Baldwin," January 23, 1886, p. 4.
40. *Los Angeles Daily Herald*, "Seventeenth and Last Day of the Great Trial," February 20, 1886, p. 5.
41. Glasscock, p. 262.
42. "E.J. Baldwin Account with Lloyd and Wood, Attorneys at Law, 1896-97."
42. *Los Angeles Times*, "Lucky Baldwin—Proud Monarch of Queerest of Little Kingdoms," November 3, 1907, Pt. II, p. 1.
44. Glasscock, pp. 272-73.
45. Glasscock, p. 272.
46. Glasscock, p. 246.
47. Glasscock, pp. 246-47.
48. *Los Angeles Times*, "Lucky on the Turf," July 24, 1899, Pt. I, p. 2.
49. Ibid.
50. *Sierra Madre Vista*, "Outdoor Industries in Southern California," October, 1883, p. 1.
51. Glasscock, p. 211.
52. *Los Angeles Examiner*, "Increase in Realty Clears Baldwin Estate," June 15, 1910, p. 11.
53. Glasscock, p. 247.
54. Glasscock, p. 249.

55. Glasscock, p. 265.
56. *Examiner*, "Increase in Realty Clears Baldwin Estate," June 15, 1910, p. 11.
57. Glasscock, p. 254.
58. Works and Lee, Attorneys at Law, letter to H.A. Unruh, May 25, 1897, Anita M. Baldwin Collection, Box 5, Rare Book Room, Huntington Library.
59. Glasscock, p. 254.
60. *Los Angeles Times*, "Baldwin's Big Hotel," November 23, 1898, Pt. I, p. 6.
61. "The Baldwin Hotel, San Francisco," *Pacific Rural Press*, XXXI (June 3, 1882), 448.
62. *Los Angeles Times*, "Smoking Ruin—Baldwin Hotel Entirely Burned Down," November 24, 1898, Pt. I, p. l.
63. *Times*, "Smoking Ruin—Baldwin Hotel Entirely Burned Down," November 24, 1898, Pt. I, p. 4. Glasscock states that Baldwin had allowed his insurance on the hotel to lapse (Glasscock, p. 273).
64. *Los Angeles Times*, "Crowding Lucky," November 26, 1898, Pt. I, p. 3.
65. Glasscock, p. 274.
66. *Los Angeles Times*, "Baldwin's Intentions," November 29, 1898, Pt. I, p. 5.
67. *Times*, "Crowding Lucky," November 26, 1898, Pt. I, p. 3.
68. Glasscock, p. 275.
69. *Los Angeles Times*, "Lucky Baldwin's Latest Fortune," March 9, 1902, Pt. IV, p. 1.
70. Glasscock, p. 281.
71. Ibid.
72. Glasscock, pp. 282-83.
73. Glasscock, pp. 283-84.
74. Glasscock, pp. 284-85. Unruh sold the supplies and equipment for a $50,000 to $60,000 profit, but was unable to sell the lot he had purchased. The land, still in Baldwin's possession when he died, was leased to the first telephone company established in Nome.
75. *Times*, "Lucky Baldwin's Latest Fortune," March 9, 1902, Pt. IV, p. 1.

CHAPTER FIVE

City-Making - Methods and Motives

1. *Los Angeles Times*, "Lucky Baldwin's Latest Fortune," March 9, 1902, Pt. IV, p. l.
2. Ibid.
3. Ibid.
4. C.B. Glasscock, *Lucky Baldwin: The Story of an Unconventional Success* (Indianapolis: Bobbs-Merrill, 1933), p. 267.
5. Los Angeles County, Hall of Records, *Subdivision Maps*, Book 5, "Lucky Baldwin Tract," p. 52.
6. H.A. Unruh, letter to C.A. Canfield, August 20, 1908.
7. H.A. Unruh, letter to C.A. Canfield, August 17, 1908.

8. Ibid.
9. Ibid.
10. Unruh, letter to Canfield, August 20, 1908.
11. Ibid.
12. Los Angeles County, Hall of Records, *Miscellaneous Records*, Book 66, "E.J. Baldwin's First Subdivision in the Rancho Potrero Grande," pp. 94-95. This property was bisected by the San Gabriel River.
13. Los Angeles County, Hall of Records, *Miscellaneous Records*, Book 70, "Map of E.J. Baldwin's Second Subdivision in Rancho Potrero Grande," p. 92. This property straddled the Potrero Grande/Felipe Lugo boundary.
14. Unruh, letter to Canfield, August 20, 1908.
15. "Inventory and Appraisement in the Matter of the Estate of Elias J. Baldwin," June 13, 1910, Anita M. Baldwin Collection, Box 6, Rare Book Room, Huntington Library.
16. Los Angeles County, Hall of Records, *Subdivision Maps*, Book 6, "El Monte Walnut Place," p. 104.
17. Donald L. Hathcock, "The History of Baldwin Park, California" (Master's thesis, Whittier College, 1954), p. 91. Covina had begun negotiations with the Pacific Electric in 1903 and by 1905 had secured a right-of-way from the Rio Hondo River at El Monte to a line just east of Covina. The right-of-way, according to Mr. Hathcock, came "almost entirely" from the lands of E.J. Baldwin.
18. Hathcock, p. 93. Vineland was known as Pleasant View in 1878 when the settlement was begun on public lands removed from Henry Dalton's Rancho Azusa de Duarte. The name was changed in the 1880s. Today's Ramona Boulevard was the dividing line between Vineland and Baldwin's La Puente acreage.
19. Hathcock, p. 93.
20. Hathcock, p. 94.
21. Gordon S. Eberly, *Arcadia—City of the Santa Anita* (Claremont: Saunders Press, 1953), p. 40.
22. *Los Angeles Times*, "Supervisors 'See' Arcadia," May 17, 1903, Pt. II, p. 1.
23. Ralph D. Paine, "Where Ranch and City Meet," *Outing*, XLVIII (August, 1906), 531.
24. Eberly, p. 48.
25. Mushet Audit Company and Bradner W. Lee, "In the Matter of the Estate of Elias J. Baldwin, Deceased. Inventory of Books, Documents, Maps, Correspondence, Miscellaneous Papers, Etc.," April 16, 1913, Anita M. Baldwin Collection, Volume 1, Rare Book Room, Huntington Library.
26. *Los Angeles Times*, "New Line Opened—Monrovia Welcomes Trolley," March 2, 1903, Pt. I, p. 11.
27. "Record of the Board of Trustees—City of Arcadia," March 15, 1903 to April 15, 1909, p. 1, Arcadia City Hall. A city of the sixth class was one with a population of less than three thousand.
28. *Los Angeles Evening Express*, "Denounced as a Fraud," May 21, 1903, p. 5.
29. Richard Bigger and James G. Kitchen, *Metropolitan Los Angeles: A Study in Integration*, Vol. II, *How the Cities Grew* (Los Angeles: Haynes Foundation, 1952), pp. 42-44.
30. *Times*, "Supervisors 'See' Arcadia, May 17, 1903, Pt. II, p. 1.
31. *Times*, "Nary a 'Nay' Has Arcadia," July 28, 1903, Pt. II, p. 1.

32. *Times*, "Supervisors 'See' Arcadia," May 17, 1903, Pt. II, p. 1.
33. *Express*, "Denounced as a Fraud," May 21, 1903, p. 5.
34. Charles W. Jones, interview by Friends of the Arcadia Library, Arcadia, California, 1963. (Transcript.)
35. *Los Angeles Times*, "Lucky Offers Bonds to Make Town Good," May 22, 1903, Pt. III, p. 12.
36. *Times*, "Supervisors 'See' Arcadia," May 17, 1903, Pt. II, p. 1.
37. Ibid.
38. *Times*, "Lucky Offers Bonds," May 22, 1903, Pt. III, p. 12.
39. *Times*, "Supervisors 'See' Arcadia," May 17, 1903, Pt. II, p. 1.
40. *Times*, "Lucky Offers Bonds," May 22, 1903, Pt. III, p. 12.
41. "Record of the Board of Trustees—City of Arcadia," p. 1.
42. *Times*, "Nary a 'Nay' Has Arcadia," July 28, 1903, Pt. II, p. 1.
43. Ibid.
44. *Pasadena Evening Star*, "Arcadia Unanimous for Incorporation," July 28, 1903, p. 5.
45. Baldwin Ranch Labor Ledger, May 1, 1901 to January 1, 1904, Anita M. Baldwin Collection, Volume III, Rare Book Room, Huntington Library.
46. *Times*, "Nary a 'Nay' Has Arcadia," July 28, 1903, Pt. II, p. 1.
47. "Record of the Board of Trustees—City of Arcadia," p. 3.
48. Ibid.
49. *Los Angeles Times*, "Lucky Baldwin—Proud Monarch of Queerest of Little Kingdoms," November 3, 1907, Pt. II, p. 1.
50. Jones.
51. *Los Angeles Evening Express*, "Lucky Will Be Its Mayor," July 27, 1903, p. 1.
52. Bigger and Kitchen, p. 158.
53. *Times*, "Lucky Baldwin Proud Monarch," November 3, 1907, Pt. II, p. 1.
54. Bigger and Kitchen, p. 157.
55. *Times*, "Lucky Offers Bonds," May 22, 1903, Pt. III, p. 12.
56. *Times*, "Nary a 'Nay' Has Arcadia," July 28, 1903, Pt. II, p. 1.
57. Ibid.
58. Carleton Burke, "Pastime of Millions," *The Thoroughbred*, IV (February, 1945), 14.
59. Glasscock, p. 291.
60. *Los Angeles Times*, "Speed Horse Arena Is Opened at Ascot," December 25, 1903, Pt. II, p. 1.
61. *Times*, "Nary a 'Nay' Has Arcadia," July 28, 1903, Pt. II, p. 1.
62. Ibid.
63. Eberly, pp. 183-84. The Baldwin Ranch school had been discontinued when Sierra Madre built a schoolhouse at Baldwin and Central Avenues in 1884-85.
64. Eberly, p. 184. The packing shed was superseded in 1907 by a two-room frame schoolhouse built on lots purchased by the district at the corner of First Avenue and California Street.
65. A search of state and local archives has uncovered only two extant issues of the *Arcadia Bulletin*:: a complete issue of Volume 1, Number 1 (May 7, 1904) is privately owned by Mrs. Vesta Tucker Reeves, a long-time area resident; excerpts from an issue of December 24, 1904 (owned by former Mayor Richard Krebs, Jr.) are among the Local History Collection of the Arcadia Public Library.

66. Eberly, pp. 57-66, 197.
67. *Arcadia Bulletin* (excerpts), December 24, 1904, Local History Collection, Arcadia Public Library.
68. "Record of the Board of Trustees—City of Arcadia," pp. 10-11.
69. "Record of the Board of Trustees—City of Arcadia," p. 17. Clara married her fourth husband, opera singer Harold Stocker, in 1882.
70. "Record of the Board of Trustees—City of Arcadia," pp. 42, 53.
71. "Sights and Scenes Along the Huntington System of Railways," *P.E. Topics*, I (June, 1906), 17.
72. Julian Fisher, personal interview, Monrovia, California, November 2, 1976.
73. John Wiley, *History of Monrovia* (Pasadena: Pasadena Star-News, 1927), p. 96.
74. Ralph D. Paine, "Where Ranch and City Meet," *Outing*, XLVIII (August, 1906), 531.
75. Paine, p. 536.
76. Pacific Electric Railway, "Monrovia and Baldwin's Ranch—The Orange Grove Route" (brochure), circa 1905, Local History Collection, Arcadia Public Library.
77. Ibid.
78. "The Baldwin Ranch" (postcard and message), circa 1905, Historical Archives, Los Angeles State and County Arboretum.
79. *Los Angeles Examiner*, "Lucky Home Again With His Sight Restored," December 31, 1905, p. 27.
80. Ibid.
81. Glasscock, pp. 294-95.
82. *Los Angeles Examiner*, "Santa Anita Opens Today; Baldwin Dream a Reality," December 7, 1907, p. 13.
83. Bigger and Kitchen, pp. 158-60.
84. Burke, p. 54.
85. *Times*, "Lucky Baldwin—Proud Monarch of Queerest of Little Kingdoms," November 3, 1907, Pt. II, p. 10.
86. Los Angeles Racing Association, "Minute Book. Organization of the Los Angeles Racing Association, September 30, 1907 to December 30, 1921," Anita M. Baldwin Collection, Box 3, Rare Book Room, Huntington Library.
87. Ibid.
88. *Examiner*, "Santa Anita Opens Today," December 7, 1907, p. 13.
89. Los Angeles Racing Association, *Souvenir, Santa Anita Park, Arcadia, Cal. Season 1908 and 1909* (Los Angeles: J.C. Effinger, 1908), California Turf Association Library, Arcadia, California.
90. Los Angeles Racing Association, "Minute Book."
91. *Los Angeles Examiner*, "Seven Thousand Turn Out to Dedicate Arcadia Track," December 8, 1907, p. 1.
92. *Examiner*, "Santa Anita Opens Today," December 7, 1907, p. 13.
93. *Los Angeles Examiner*, "How to Reach Santa Anita," December 7, 1907, p. 1.
94. *Los Angeles Evening Express*, "Santa Anita Race Track Now Is Open," December 7, 1907, p. 2.
95. *Examiner*, "Seven Thousand Turn Out to Dedicate Arcadia Track," December 8, 1907, p. 1.
96. *Los Angeles Times*, "Fortunes Are Lost or Won at Arcadia," December 8, 1907, Pt. VIII, p. 1.
97. Glasscock, p. 295.
98. *Los Angeles Times*, "California's Last Race Stirs With Its Humor and Tragedy,"

April 18, 1909, Pt. II, p. 3.
99. *Express*, "Santa Anita Race Track Now Is Open," December 7, 1907, p. 2.
100. *Los Angeles Times*, "Sunday Face for Arcadia," March 31, 1909, Pt. II, p. 1.
101. *Los Angeles Times*, "How Did It Happen? Old Lucky's Historic Oakwood Lies in Ashes," November 27, 1911, Pt. II, p. 1. Estate records indicate that the Oakwood was leased to Lawrence and Comstock for $500 per month (Hull McClaughry, "Report to Mrs. Anita Baldwin McClaughry," October 31, 1910, Anita M. Baldwin Collection, Box 6, Rare Book Room, Huntington Library).
102. Fisher, personal interview, November 2, 1976.
103. Letters from Dave Coapman, boxing researcher, Arcadia, California, 1978-79.
104. Vesta Tucker Reeves, telephone interview, Rosemead, California, August 9, 1979.
105. Julian Fisher, personal interview, Monrovia, California, August 4, 1976.
106. Eberly, p. 44. Estate records indicate that the "Norfolk" was leased by Baldwin for $100 per month (Mushet Audit Company and Bradner W. Lee, "Inventory of Books, Documents, Maps, Correspondence, Miscellaneous Papers, Etc.," April 16, 1913).
107. Reeves.
108. Los Angeles Racing Association, *Souvenir, Santa Anita Park, Arcadia, Cal. Season 1908 and 1909.*

CHAPTER SIX

Death and Estate Settlement

1. *Los Angeles Times*, "Lucky Baldwin—Proud Monarch of Queerest of Little Kingdoms," November 3, 1907, Pt. II, p. 1.
2. *Los Angeles Examiner*, "Lucky Baldwin Dead; Fortune Rated Up to $25 Million," March 2, 1909, p. 2.
3. *Los Angeles Times*, "His Greatest Race Run, 'Lucky' Loses," March 2, 1909, Pt. I, p. 1.
4. *Los Angeles Examiner*, "Classical Music To Be Only Funeral Service," March 3, 1909, p. 3.
5. *Los Angeles Evening Express*, "Racetrack Will Close Saturday," April 7, 1909, p. 1.
6. *Los Angeles Times*, "Sunday Face for Arcadia," March 31, 1909, Pt. II, p. 1.
7. *Los Angeles Times*, "California's Last Race Stirs With Its Humor and Tragedy," April 18, 1909, Pt. II, p. 1.
8. Gordon S. Eberly, *Arcadia—City of the Santa Anita* (Claremont: Saunders Press, 1953), p. 56.
9. *Los Angeles Times*, "How Did It Happen? Old Lucky's Historic Oakwood Lies in Ashes," November 27, 1911, Pt. II, p. 1.
10. Eberly, pp. 60-62, 71.
11. Eberly, p. 222.
12. *Los Angeles Times*, "Lucky's Millions Cut to Measure," March 4, 1909, Pt. I, p. 14.
13. Ibid.
14. Elias J. Baldwin, Last Will and Testament, November 4, 1908, Special Collections, University of California, Los Angeles.
15. *Los Angeles Times*, "Baldwin Heiress Mystery Deepens," March 6, 1909, Pt. II, p. 2. Adding to

the mystery was a statement by Police Sergeant Abraham Sharpe that Zelda Selby was his granddaughter. Sergeant Sharpe, it was reported, was Lucky Baldwin's "closest friend." The sergeant made no mention of Martha Agnes Fowler; *Los Angeles Examiner*, "Baldwin Will Owns Mysterious Heir as Daughter," March 4, 1909, p. 3. The *Examiner* made no mention of Sergeant Sharpe and reported that Zelda Selby had been raised by her grandmother, "Mrs. Robinson."

16. Elias J. Baldwin, Last Will and Testament, November 4, 1908.
17. "Inventory and Appraisement in the Matter of the Estate of Elias J. Baldwin," June 13, 1910, Anita M. Baldwin Collection, Box 6, Rare Book Room, Huntington Library.
18. Elias J. Baldwin, Last Will and Testament, November 4, 1908.
19. *Times*, "Lucky's Millions Cut to Measure," March 4, 1909, Pt. I, p. 14.
20. "Inventory and Appraisement," June 13, 1910. Real property, $10,611,025; personal property, $318,776.
21. Hull McClaughry, "Report to Mrs. Anita Baldwin McClaughry," October 31, 1910, Anita M. Baldwin Collection, Box 6, Rare Book Room, Huntington Library. Baldwin mentioned the then "pending" mortgage in Article 16 of his will (Elias J. Baldwin, Last Will and Testament, November 4, 1908).
22. McClaughry. Notations of Unruh's payments appear in his court reports (Estate Papers of Elias Jackson Baldwin, "Report to Superior Court, County of Los Angeles, by Hiram A. Unruh, Executor," Report 1, April 9, 1909 to March 31, 1910, Anita M. Baldwin Collection, Box 6, Rare Book Room, Huntington Library).
23. California Scrapbook—Biography (collection of undated newspaper clippings), *Los Angeles Examiner*, "Astonishing Settlement," circa 1910, California Room, Los Angeles Public Library.
24. Clara's two children were Albert Snyder, Jr., by her second husband, and Rosebudd Doble Mullender, by her third husband. Anita's two children by her second husband, Hull McClaughry, were Dextra and Baldwin McClaughry. When Anita divorced McClaughry in 1915, she legally resumed her maiden name and had her children's names changed to Dextra and Baldwin M. Baldwin.
25. Elias J. Baldwin, Last Will and Testament, November 4, 1908. In 1936, Anita Baldwin sold the then remaining thirteen hundred acres surrounding the homesite to Harry Chandler's real estate corporation, and in 1947 Chandler removed 111 acres of the property from his Arcadia subdivisions and sold the land jointly to the State of California and the County of Los Angeles for use as a public garden.
26. *Los Angeles Examiner*, "$20 Million for Two Baldwin Heirs," April 2, 1912, p. 2. The La Cienega property produced oil profits for the daughters far in excess of $5 million in subsequent years until it, too, was sold for subdivision.
27. California Scrapbook—Biography, *Examiner*, "Astonishing Settlement," circa 1910.
28. *Los Angeles Examiner*, "Increase in Realty Clears Baldwin Estate," June 15, 1910, p. 11.
29. California Scrapbook—Biography, *Examiner*, "Astonishing Settlement," circa 1910.
30. *Examiner*, "Increase in Realty Clears Baldwin Estate," June 15, 1910, p. 11.
31. *Los Angeles Times*, "United for Battle: Widow Must Defend Old Lucky's Will," September 25, 1910, Pt. II, p. 1.

32. Ibid.
33. Ibid.
34. Glasscock, pp. 305-306.
35. California Scrapbook—Biography (collection of undated newspaper clippings), "Twenty Million Dollars Realized for Heirs of Late E.J. Baldwin," circa 1913, California Room, Los Angeles Public Library.
36. *Los Angeles Times*, "Four Years of Toil Crowned," April 25, 1913, Pt. II, p. 1.
37. California Scrapbook—Biography, "20 Million Realized for Heirs of Late E.J. Baldwin," circa 1913.
38. Ibid.
39. *Los Angeles Examiner*, "Baldwin Estate Is Settled," April 25, 1913, p. 19.

Summary

1. *Los Angeles Examiner*, "Lucky Baldwin Dead; Fortune Rated Up to $25 Million," March 2, 1909, p. 1.
2. *Los Angeles Evening Express*, "Splendid Summer House," March 20, 1875, p. 3.
3. *Los Angeles Daily Herald*, "Baldwin's Life One of Adventure," March 2, 1909, Pt. II, p. 2.
4. *Los Angeles Times*, "Lucky Baldwin—A Look at His Famous Ranch and Its Wonders," June 11, 1893, p. 9.
5. Hubert Howe Bancroft, *Chronicles of the Builders of the Commonwealth*, III (San Francisco: The History Company, 1892), p. 369.
6. Ralph D. Paine, "Where Ranch and City Meet," *Outing*, XLVIII (August, 1906), 536.
7. Bancroft, III, p. 364.
8. *Herald*, "Baldwin's Life One of Adventure," March 2, 1909, Pt. II, p. 1.
9. C.B. Glasscock, *Lucky Baldwin: The Story of an Unconventional Success* (Indianapolis: Bobbs-Merrill, 1933), p. 300.
10. *Los Angeles Times*, "Baldwin's Big Mortgage," January 24, 1896, Pt. I, p. 6.
11. Glasscock, p. 308.

BIBLIOGRAPHY

Books

Bancroft, Hubert Howe. *Chronicles of the Builders of the Commonwealth.* 7 vols. San Francisco: The History Company, 1892.

Beckwith, B.K. *The Story of Santa Anita—The Great Race Place.* Los Angeles: Day Printing Corp., 1975.

Bigger, Richard, and James D. Kitchen. *Metropolitan Los Angeles: A Study in Integration.* Vol. II, *How the Cities Grew.* Los Angeles: Haynes Foundation, 1952.

Bowen, Edith, ed. *Annals of Early Sierra Madre.* Sierra Madre: Sierra Madre Historical Society, 1950.

Carew, Harold. *History of Pasadena and the San Gabriel Valley.* Los Angeles: S.J. Clarke Publishing Company, 1930.

Colver, Seth, ed. *The Historical Volume and Reference Works: Covering Arcadia, Azusa, Baldwin Park, Bradbury, Charter Oak, Claremont, Covina, Duarte, El Monte, Glendora, Irwindale, La Verne, Monrovia, Pomona, San Dimas, South El Monte, West Covina.* Arlington, California: Historical Publishers, 1964.

Dumke, Glenn S. *The Boom of the Eighties in Southern California.* San Marino: Huntington Library, 1944.

Eberly, Gordon S. *Arcadia—City of the Santa Anita.* Claremont: Saunders Press, 1953.

French, Virginia. *Rancho La Cienega O Paso de la Tijera.* Los Angeles: John D. Roche, Inc., 1970.

Glasscock, C.B. *Lucky Baldwin: The Story of an Unconventional Success.* Indianapolis: Bobbs-Merrill, 1933.

Graves, Jackson. *California Memories,* 1857-1930. Los Angeles: Times Mirror Press, 1930.

Grenier, Judson A., ed. *A Guide to Historic Places in Los Angeles County.* Dubuque, Iowa: Kendall/Hunt Publishing Company, 1978.

Guinn, James M. *A History of California and an Extended History of Its Southern Coast Counties.* 3 vols. Los Angeles: Historic Record Company, 1907.

Hall, William Hammond. *Irrigation in Southern California.* Sacramento: State Printing Office, 1888.

An Illustrated History of Los Angeles County, California. Chicago: Lewis Publishing Company, 1889.

Lindley, Walter, and J.P. Widney. *California of the South.* New York: D. Appleton and Company, 1888.

Lyman, George. *Ralston's Ring: California Plunders the Comstock.* New York: Charles Scribner's Sons, 1937.

McWilliams, Cary. *Southern California: An Island on the Land.* Santa Barbara: Peregrine Smith, Inc., 1946.

Mendenhall, Walter C. *Ground Waters and Irrigation Enterprises in the Foothill Belt, Southern California..* Washington: Government Printing Office, 1908.

Nadeau, Remi. *City-Makers: The Men Who Transformed Los Angeles From Village to Metropolis During the First Great Boom, 1868-76.* Garden City: Doubleday, 1948.
Newmark, Maurice H., and Marco R. Newmark, eds. *Sixty Years in Southern California, 1853-1913. Containing the Reminiscences of Harris Newmark.* 3d ed. Boston and New York: Houghton Mifflin, 1930.
Pacific Coast Annual Mining Review and Stock Ledger. San Francisco: Frances and Valentine, 1878.
Pflueger, Donald H. *Covina: Sunflowers, Citrus, Subdivision.* Pasadena: Castle Press, 1964.
Radford, Claire. *La Puente Kaleidoscope.* La Puente: La Puente Valley Woman's Club, 1974.
Reid, Dr. Hiram A. *History of Pasadena.* Pasadena: Pasadena History Company, Publishers, 1895.
Robinson, W.W. *Ranchos Become Cities.* Pasadena: San Pasqual Press, 1939.
Scott, Edward B. *The Saga of Lake Tahoe.* Lake Tahoe: Sierra-Tahoe Publishing Co., 1957.
Shuck, Oscar. *Historical Abstract of San Francisco.* 3 vols. San Francisco: Oscar Shuck, 1897.
Studer, Robert, ed. *The Historical Volume and Reference Works: Including Alhambra, Monterey Park, Rosemead, San Gabriel, Temple City.* Los Angeles: Historical Publishers, 1962.
Swain, John G., ed. *The Historical Volume and Reference Works: Covering Bassett, City of Industry, La Habra, La Mirada, La Puente, Pico Rivera, Santa Fe Springs, Walnut, Whittier.* Whittier: Historical Publishers, 1963.
Wait, Frona. *Wines and Vines of California: A Treatise on the Ethics of Wine-Drinking.* San Francisco: Bancroft Company, 1889.
Wiley, John. *History of Monrovia.* Pasadena: Pasadena Star-News, 1927.

Periodicals

"Baldwin, Elias Jackson." *The National Cyclopedia of American Biography*, XXII (1932), 381-82.
"The Baldwin Hotel, San Francisco." *The Pacific Rural Press*, XXXI (June 3, 1882), 448
"A Big Rancho." *Rural Californian*, IX (November, 1886), 244.
Burke, Carleton. "Pastime of Millions." *The Thoroughbred*, IV (February, 1945), 14-15, 54.
______. "Pastime of Millions." *The Thoroughbred*, IV (April-May, 1945), 16-17, 32.
Hamblin, Judge Donald William. "Lucky's Litigious Loves." *The Branding Iron*, March, 1969, pp. 1, 4-10.
Hoyt, Franklyn. "The San Gabriel Valley Rapid Transit Railroad." *Southern California Quarterly*, XXXIII (September, 1951), 213-20.
Netz, Joseph. "The Great Los Angeles Real Estate Boom of 1887." *Annual Publications—Historical Society of Southern California*, X (1915-16), 54-68.
Paine, Ralph D. "Where Ranch and City Meet." *Outing*, XLVIII (August, 1906), 528-41.
Payne, Gerald. "Millionaire, King of a Thousand Queens." *True*, February, 1957, pp. 21-33, 104.
"Prince of the House of Argonauts." *San Gabriel Valley Digest*, III (April, 1930), 6-7, 29.
Robinson, John W. "A Mountain of Gold." *Desert*, XXXVIII (September, 1975), 34-37.
"Sights and Scenes Along the Huntington System of Railways." *P. E. Topics*, I (June, 1906), 16-17.
Slosson, H.V. "The Orange Interest of San Gabriel Valley; Vineyards of San Gabriel Valley." *The Pacific Rural Press*, XIX (February 28, 1880), 136-37.
Stull, Henry. "A Day at the 'Home of Grinstead.'" *Spirit of the Times*, CXXXVII

(March 18, 1899), 861-63.

Newspapers

Arcadia Bulletin, May 7, 1904. Privately owned by Mrs. Vesta Tucker Reeves.
_______. Excerpts, December 24, 1904. Local History Collection, Arcadia Public Library.
The Illustrated Los Angeles Herald. "The Santa Anita Tract," April, 1886.
_______. "The Race Horse in Los Angeles," June, 1889.
_______. "A Rich Possession—The Home of E.J. Baldwin, Esq., Los Angeles County," June, 1889.
_______. "San Gabriel Valley—Choice Lands Offered For Sale by E.J. Baldwin" (advertisement), July, 1893.
Los Angeles Daily Herald. "Santa Anita," March 21, 1875.
_______. "Local Brevities," October, 1875.
_______. "San Gabriel Fruit Belt," October 24, 1875.
_______. "An Unwarranted Attack," January 19, 1876.
_______. "Mustering of Creditors of Temple and Workman," February 5, 1876.
_______. "Lucky Baldwin and His Ranch," February 6, 1876.
_______. "The Santa Anita Tract on Sale in Five to Thirty Acre Lots" (advertisement), January 6, 1886.
_______. "Local News Notes," January 17, 1886.
_______. "Railroad News Notes," January 17, 1886.
_______. "Breach of Promise Case," February 4, 1886.
_______. "Seventeenth and Last Day of the Great Trial," February 20, 1886.
_______. "Local News Notes," March 23, 1886.
_______. "The Work of Development in the Foothills," June 1, 1886.
_______. "Santa Anita Station," June 13, 1886.
_______. "Monrovia Station—Santa Anita Tract" (advertisement), November 16, 1886.
_______. "The Orange Belt Route" (advertisement), November 21, 1886.
_______. "Monrovia Movements," December 9, 1886.
_______. "Railway Systems," December 9, 1886.
_______. "L.A. and S.G.V.R'y," January 1, 1887.
_______. "The New Railroad," January 18, 1887.
_______. "Felipe Lugo Tract" (advertisement), January 23, 1887.
_______. "Santa Anita Tract—Part of the Celebrated Santa Anita Ranch, Now On Sale" (advertisement), January 23, 1887.
_______. "Perkins—Baldwin," March 15, 1887.
_______. "Beautiful Arcadia," March 30, 1887.
_______. "Santa Anita Stallions" (advertisement), April 1, 1887.
_______. "Arcadia" (advertisements), April 1 to June 9, 1887.
_______. "News Notes," April 3, 1887.
_______. "Arcadia," April 9, 1887.
_______. "Arcadia" (advertisement), April 12, 1887.

_______. "Arcadia," April 18, 1887.
_______. "A New Railroad," April 22, 1887.
_______. "Arcadia," April 28, 1887.
_______. "Arcadia," May 5, 1887.
_______."Arcadia," May 21, 1887.
_______. "News Notes," June 7, 1887.
_______. "Baldwin Brings an Action For Considerable Water," July 1, 1887.
_______. "Another Railroad," July 9, 1887.
_______. "Rapid Transit," July 10, 1887.
_______. "L.A. and S.G.V.R.T.R.R.," July 21, 1887.
_______. "The Perkins Case," July 26, 1887.
_______. "S.G.V. Rapid Transit R'y" (advertisement), January 21, 1889.
_______. "Baldwin's Life One of Adventure," March 2, 1909.
_______. "Death of Lucky Baldwin Closes Famous Career," March 2, 1909.
_______. "Baldwin Gives Widow Only a Trifling Share," March 4, 1909.
Los Angeles Evening Express. "What Lucky Baldwin Thinks," February 15, 1875.
_______. "Gold Mountain," March 19, 1875.
_______. "Local Items," March 20, 1875.
_______. "Splendid Summer House," March 20, 1875.
_______. "Most Important Transaction of the Day," March 22, 1875.
_______. "How It Works," March 25, 1875.
_______. "About the Adjective 'Lucky,'" March 27, 1875.
_______. "The Santa Anita Ranch," October 14, 1876.
_______. "The Perils of Riches," April 24, 1877.
_______. "Local Dots," May 6, 1877.
_______. "The Baldwin Distillery Case," February 6, 1880.
_______. "Denounced as a Fraud," May 21, 1903.
_______. "Lucky Will Be Its Mayor," July 27, 1903.
_______. "Only Thirty-Eight Votes," July 28, 1903.
_______. "Santa Anita Race Track Now Is Open," December 7, 1907.
_______. "Lucky Baldwin Is Dead," March 1, 1909.
_______. "Racetrack Will Close Saturday," April 7, 1909.
Los Angeles Examiner. "Lucky Home Again With His Sight Restored," December 31, 1905.
_______. "How to Reach Santa Anita," December 7, 1907.
_______. "Santa Anita Opens Today; Baldwin Dream a Reality," December 7, 1907.
_______. "Seven Thousand Turn Out to Dedicate Arcadia Track," December 8, 1907.
_______. "Baldwin's Assessments in This County, $6,545,869," March 2, 1909.
_______. "Lucky Baldwin Dead: Fortune Rated Up to $25 Million," March 2, 1909.
_______. "Baldwin Will Satisfied Heirs—Pretty Nurse Is Provided For," March 3, 1909.
_______. "Classical Music to Be Only Funeral Service," March 3, 1909.
_______. "Baldwin Will Owns Mysterious Heir As Daughter," March 4, 1909.
_______. "Increase in Realty Clears Baldwin Estate," June 15, 1910.
_______. "$20 Million for Two Baldwin Heirs," April 2, 1912.

_______. "Baldwin Estate Is Settled," April 25, 1913.
Los Angeles Times. "A New Colony—Monrovia," January 1, 1886.
_______. "Baldwin vs. Bell," January 15, 1886.
_______. "Angry Baldwin," January 23, 1886.
_______. "Railroads," April 22, 1886.
_______. "Baldwin's Belvedere," May 8, 1886.
_______. "Railroad News," October 15, 1886.
_______. "The Santa Fe's Enormous Deal Near the Raymond," April 7, 1887.
_______. "The New System," April 27, 1887.
_______. "The Railways—The Monrovia Dummy Line," May 5, 1887.
_______. "Railways—The Three Best," May 5, 1887.
_______. "Incorporated," May 8, 1887.
_______. "Sporting News," February 2, 1891.
_______. "E.J. Baldwin's Vast Irrigating Enterprise," August 20, 1891.
_______. "Pasadena and Thereabouts," December 4, 1891.
_______. "Lucky Baldwin—A Look at His Famous Ranch and Its Wonders," June 11, 1893.
_______. "Lucky Baldwin," January 28, 1895.
_______. "Baldwin's Big Mortgage," January 24, 1896.
_______. "Baldwin's Big Hotel," November 23, 1898.
_______. "Smoking Ruin—Baldwin Hotel Entirely Burned Down," November 24, 1898.
_______. "Crowding Lucky," November 26, 1898.
_______. "Baldwin's Intentions," November 29, 1898.
_______. "Lucky on the Turf," July 24, 1899.
_______. "Hotel Site Sold," October 29, 1899.
_______. "Lucky Baldwin's Latest Fortune," March 9, 1902.
_______. "New Line Opened—Monrovia Welcomes Trolley," March 2, 1903.
_______. "Supervisors 'See' Arcadia," May 17, 1903.
_______. "Lucky Offers Bonds to Make Town Good," May 22, 1903.
_______. "Nary a 'Nay' Has Arcadia," July 28, 1903.
_______. "Work at Ascot Begins in Earnest," July 31, 1903.
_______. "Speed Horses Arena Is Opened at Ascot," December 25, 1903.
_______. "Lucky Baldwin Proud Monarch of Queerest of Little Kingdoms," November 3, 1907.
_______. "Fortunes Are Lost or Won at Arcadia," December 8, 1907.
_______. "Famous Racers Lucky Owned," March 2, 1909.
_______. "His Girl Wife Had Patience," March 2, 1909.
_______. "His Greatest Race Run, 'Lucky' Loses," March 2, 1909.
_______. "Lucky's Millions Cut to Measure," March 4, 1909.
_______. "Baldwin Heiress Mystery Deepens," March 6, 1909.
_______. "Two Sides of Old Lucky," March 21, 1909.
_______. "Sunday Face for Arcadia," March 31, 1909.
_______. "California's Last Race Stirs with Its Humor and Tragedy," April 18, 1909.
_______. "United for Battle: Widow Must Defend Old Lucky's Will," September 25, 1910.
_______. "How Did It Happen? Old Lucky's Historic Oakwood Lies in Ashes," November 27, 1911.

_______. "Millions the Stake Here," November 28, 1912.
_______. "Baldwin Millions To Be Distributed Today," April 24, 1913.
_______. "Four Years of Toil Crowned," April 25, 1913.
_______. "Busy Life Is Snuffed Out," December 17, 1916.
Monrovia Messenger. "Formal Opening," August 30, 1888.
_______. "Hotel Oakwood," September 20, 1888.
_______. "Santa Anita War," November 22, 1888.
_______. "To Rent," December 13, 1888.
_______. "An Explanation," May 2, 1889.
_______. "The Hotel Oakwood: Arcadia's Pretty Hostelry Open for the Winter," December 26, 1889.
_______. "Baldwin's Brood Mares," March 20, 1890.
_______. "Santa Anita Ranch," January 22, 1891.
_______. "Arcadia and Santa Anita," February 12, 1891.
_______. "The Baldwin Tract," March 12, 1891.
Monrovia Planet. "The Baldwin Stable," December 4, 1886.
_______. "A Local Railroad—Orange Belt Route Through the San Gabriel Valley," January 8, 1887.
_______. "Santa Anita Tract" (advertisement), January 8, 1887.
_______. "The New Railway," January 22, 1887.
_______. "Another Railroad—The Southern Pacific Locating a Line Through Monrovia," February 26, 1887.
_______. "A Perfume Factory," March 5, 1887.
_______. "A New Town," March 12, 1887.
_______. "Monrovia's Future Railroads," March 19, 1887.
_______. "Satellites," March 26, 1887.
_______. "Arcadia—Monrovia's Youngest Sister and Her Progress," April 9, 1887.
_______. "Satellites," April 9, 1887.
_______. "Satellites," April 23, 1887.
_______. "Arcadia Notes," April 30, 1887.
_______. "Baldwin's Stables," April 30, 1887.
_______. "Monrovia—The Enormous Growth of a One Year Old Town," May 7, 1887.
_______. "Arcadia Notes," May 21, 1887.
_______. "Things of Beauty—A Visit to Arcadia's Beautiful Flower Farm," June 4, 1887.
_______. "Arcadia," September 24, 1887.
_______. "Regarding Incorporation," September 24, 1887.
Pasadena Evening Star. "Arcadia Unanimous for Incorporation," July 28, 1903.
Pasadena Star News. "Building of Original Ranch Units Remembered" (clipping), June 10, 1956.
San Gabriel Valley News. "El Monte," April 11, 1878.
_______. "Santa Anita," May 9, 1878.
Sierra Madre Vista. "Santa Anita Ranch," July, 1881.
_______. "Sierra Madre Tract," July, 1881.
_______. "The Baldwin," October, 1883.
_______. "Outdoor Industries in Southern California," October, 1883.

_______. "Sierra Madre and Santa Anita Tracts," October, 1883.
_______. "Vineyard, Orchard, and Farming Lands for Sale in the Santa Anita Tract" (map and advertisement), October, 1883.
_______. "Beautiful Sierra Madre—A Brief History of Its Settlement and Steady Development to Date," March 17, 1888.
_______. "Local News and Notes," March 24, 1888.
_______. "Local News and Notes," April 14, 1888.
_______. "About a Depot Building," March 9, 1889.
_______. "Hotel Oakwood" (advertisement), March 9, 1889.
_______. "S.G.V.R.T.R. Schedule" (advertisement), May 17, 1889.
_______. "Local News and Notes," August 2, 1889.
_______. "Local Views," July 3, 1890.
_______. "Santa Anita Store" (advertisement), July 3, 1890.
_______. "Local Views," August 21, 1890.
_______. "Local Views," August 28, 1890.
_______. "Local Views," October 2, 1890.
_______. "Baldwin's Racers," October 9, 1890.
_______. "Santa Anita and Mr. Baldwin," November 6, 1890.
_______. "Local Views," November 13, 1890.
_______. "Arcadia and Santa Anita," January 8, 1891.
_______. "Notes and News," January 15, 1891.
_______. "Railroad Time Table," May 21, 1891.
_______. "Local Views," May 28, 1891.
_______. "Local Views," July 23, 1891.
_______. "Local Views," August 27, 1891.
The Valley Vista. "Lucky Baldwin's Energy," March 31, 1900.

Maps and Deeds

"Birdseye View of Arcadia and Santa Anita Tract" (advertising map), circa 1887.
Cleminson, Elizabeth. "Memory of E.J. Baldwin Ranch, 1889-1890" (hand-drawn map). Historical Archives, Los Angeles State and County Arboretum.
Los Angeles County. Hall of Records. *Deeds*, Book 34, pp. 253-56.
_______. *Deeds*, Book 38, pp. 229-30.
_______. *Deeds*, Book 39, pp. 404-406.
_______. *Deeds*, Book 40, pp. 212-14, 232-35.
_______. *Deeds*, Book 42, pp. 74-76.
_______. *Deeds*, Book 76, pp. 288, 304-306.
_______. *Deeds*, Book 77, pp. 166-67.
_______. *Deeds*, Book 78, pp. 545-48.
_______. *Deeds*, Book 150, pp. 103-105.
_______. *Deeds*, Book 151, pp. 547-49.

_______. *Deeds*, Book 173, pp. 493-95.
_______. *Deeds*, Book 185, pp. 560-63.
_______. *Miscellaneous Records*, Book 42, "Map of Santa Anita Colony," p. 87.
_______. *Miscellaneous Records*, Book 42, "Map of the Western Two-Thirds of Rancho San Francisquito," pp. 93-94.
_______. *Miscellaneous Records*, Book 52, "E.J. Baldwin's Addition No. 1 to Santa Anita Colony," p. 60.
_______. *Miscellaneous Records*, Book 53, "E.J. Baldwin's Addition No. 2 to Santa Anita Colony," p. 4.
_______. *Miscellaneous Records*, Book 66, "E.J. Baldwin's First Subdivision in the Rancho Potrero Grande," pp. 94-95.
_______. *Miscellaneous Records*, Book 70, "E.J. Baldwin's Second Subdivision in Rancho Potrero Grande," p. 92.
_______. *Mortgages*, Book 17, pp. 50-55.
_______. *Subdivision Maps*, Book 5, "Lucky Baldwin Tract," p. 52.
_______. *Subdivision Maps*, Book 6, "El Monte Walnut Place," p. 104.
_______. *Subdivision Maps*, Book 7, "E.J. Baldwin's Second Subdivision of a Portion of Rancho La Puente," p. 7.
_______. *Subdivision Maps*, Book 8, "E.J. Baldwin's Fourth Subdivision of Part of Rancho La Puente," p. 186.
_______. *Subdivision Maps*, Book 10, "Arcadia Acreage Tract," p. 18.
_______. *Subdivision Maps*, Book 12, "E.J. Baldwin's Fifth Subdivision of a Portion of Rancho La Puente," p. 134.
"Map of a Part of Arcadia, 1887." Rare Book Room, Huntington Library.
"Map of a Part of Santa Anita Tract, September, 1886." Rare Book Room, Huntington Library.
"Map of Los Angeles County, 1884." Rare Book Room, Huntington Library.
"Map of the Town of Baldwin in the Santa Anita Tract, May, 1886." Rare Book Room, Huntington Library.
"Map Showing the Subdivisions of the Rancho Potrero de Felipe Lugo, September, 1886." Rare Book Room, Huntington Library.
Sierra Madre Vista. "Map of the Santa Anita Rancho—Showing E.J. Baldwin's Home Place 'Santa Anita,' Sierra Madre and Santa Anita Tract" (advertisement), October, 1883.

Other Sources

Abstract of Title and Certificate of Search to Proceedings in the Case of Camilo Martin vs. William Temple et al., April 21, 1888. Anita M. Baldwin Collection, Box 6. Rare Book Room, Huntington Library.
The Atchison, Topeka, and Santa Fe Railway Company. Personal correspondence between Thomas J. Fitzgerald, General Attorney, California, and the writer. Arcadia, California, July 2, 1979.
Baldwin, Anita. Birth Certificate, January 10, 1876. Anita M. Baldwin Collection, Box 4. Rare Book Room, Huntington Library.
Baldwin, Elias J. Last Will and Testament, November 4, 1908. Special Collections,

University of California, Los Angeles.
"The Baldwin Ranch." Postcard and message, circa 1905. Historical Archives, Los Angeles State and County Arboretum.
Baldwin Ranch Labor Ledger, May 1, 1901 to January 1, 1904. Anita M. Baldwin Collection, Volume III. Rare Book Room, Huntington Library.
Baldwin Ranch Store Accounts, 1895-97. Anita M. Baldwin Collection, Box 5. Rare Book Room, Huntington Library.
"Baldwin's Ranch. Winery and Price List. M. Lawrence and Co., Sole Agents." Pamphlet. Research Library, Los Angeles State and County Arboretum.
Butler, Samuel. Interview by Elizabeth Dixon. Oral History Program, University of California, Los Angeles, 1966. (Transcript.)
"By-Laws of Santa Anita Water Company, Los Angeles County, California—Santa Anita Tract." Rare Book Room, Huntington Library.
California Scrapbook—Biography. Collection of undated newspaper clippings. *Los Angeles Examiner*, "Astonishing Settlement." California Room, Los Angeles Public Library.
_______. $20 Million Realized for Heirs of Late E.J. Baldwin."
_______. "Woman is Loser in $2.6 Million Estate Suit."
"California's Choicest Locality—San Gabriel Valley." Brochure, 1891-92. Local History Collection, Arcadia Public Library.
Cleminson, Elizabeth. Interview by Barbara Fincher. Temple City, California, October, 1976.
Coapman, Dave. Personal correspondence between Dave Coapman, boxing researcher, and the writer. Arcadia, California, 1978-79.
"E.J. Baldwin Account with Lloyd and Wood, Attorneys at Law, 1896-97." Anita M. Baldwin Collection, Box 5, Rare Book Room, Huntington Library.
Estate Papers of Elias Jackson Baldwin. "Report to Superior Court, County of Los Angeles, by Hiram A. Unruh, Executor," Report 1, April 9, 1909 to March 31, 1910. Anita M. Baldwin Collection, Box 6. Rare Book Room, Huntington Library.
Fisher, Julian. Personal interviews. Monrovia, California, August 4, August 17, and November 2, 1976.
Fraser. "E.J. Baldwin—Notes Taken During a Visit to His Santa Anita Rancho." Dictations and materials for H.H. Bancroft's *Chronicles of the Builders of the Commonwealth*, August 26, 1891. Bancroft Library, University of California, Berkeley. (Transcript.)
Hathcock, Donald L. "The History of Baldwin Park, California." Master's thesis, Whittier College, 1954.
"How to Get a Deed from E.J. Baldwin—Lands in the San Gabriel Valley Offered for Sale by E.J. Baldwin." Brochure, 1891-92. Local History Collection, Arcadia Public Library.
"Inventory and Appraisement in the Matter of the Estate of Elias J. Baldwin," June 13, 1910. Anita M. Baldwin Collection, Box 6. Rare Book Room, Huntington Library.
Jones, Charles W. Interview by Friends of the Arcadia Library. Arcadia, California, 1963. (Transcript.)
Knisley, Ray. Personal correspondence between Ray Knisley, former Baldwin Ranch manager, and Jack McCaskill. Los Angeles State and County Arboretum, January, 1968.
Lee, Bradner W. Papers. Special Collections, University of California, Los Angeles.

The Los Angeles and San Gabriel Railway. "Prospectus," circa 1884. Rare Book Room, Huntington Library.

Los Angeles and San Gabriel Valley Rail Road Company. Contract with E.J. Baldwin, November 24, 1885. Historical Archives, Los Angeles State and County Arboretum. (Photocopy.)

Los Angeles Racing Association. "Minute Book. Organization of the Los Angeles Racing Association, September 30, 1907 to December 30, 1921." Anita M. Baldwin Collection, Box 3. Rare Book Room, Huntington Library.

_______. "Santa Anita Park 1907-08. Official Program, Saturday, December 7, 1907." Rare Book Room, Huntington Library.

_______. *Souvenir, Santa Anita Park, Arcadia, Cal. Season 1908 and 1909.* Los Angeles: J.C. Effinger, 1908. California Turf Association Library, Arcadia, California.

McCaslin, Lowry. Personal interview. Arcadia, California, July 2, 1979.

McClaughry, Hull. "Report to Mrs. Anita Baldwin McClaughry," October 31, 1910. Anita M. Baldwin Collection, Box 6. Rare Book Room, Huntington Library.

Mushet Audit Company, and Bradner W. Lee. "In the Matter of the Estate of Elias J. Baldwin, Deceased. Inventory of Books, Documents, Maps, Correspondence, Miscellaneous Papers, Etc.," April 16, 1913. Anita M. Baldwin Collection, Volume I. Rare Book Room, Huntington Library.

Pacific Electric Railway. "Monrovia and Baldwin's Ranch—The Orange Grove Route." Brochure, circa 1905. Local History Collection, Arcadia Public Library.

Pearle, Alice Warren. "Santa Anita Rancho: Its History and Personalities." Master's thesis, University of Southern California, 1939.

"Pocket Time Table—California Central Ry. and California Southern R.R." Brochure, October 7, 1888. Rare Book Room, Huntington Library.

"Record of the Board of Trustees—City of Arcadia," March 15, 1903 to April 15, 1909. Arcadia City Hall.

Reeves, Vesta Tucker. Telephone interview. Rosemead, California, August 9, 1979.

Saunders, Karen. "The Story of Lucky Baldwin." Term paper, University of California, Los Angeles, Extension, March 16, 1970.

Shoop, C.F. "History of City's First Railroad Line Traced." Newspaper clipping. Vertical File, Los Angeles State and County Arboretum.

"Sierra Madre and Santa Anita." *The Pacific Rural Press.* Magazine clipping, May 9, 1885. Historical Archives, Los Angeles State and County Arboretum.

Unruh, H.A. Letter to C.A. Canfield, August 17, 1908. Anita M. Baldwin Collection, Box 5. Rare Book Room, Huntington Library.

_______. Letter to C.A. Canfield, August 10, 1908.

Works and Lee, Attorneys at Law. Letter to H.A. Unruh, May 25, 1897. Anita M. Baldwin Collection, Box 5. Rare Book Room, Huntington Library.